INSIDE THE ANIMAL MIND

PAMELA WEINTRAUB

CENTENNIAL BOOKS

INSIDE
THE
ANIMAL
MIND

Contents

160

210

The amazing octopus can change the color of its body in just three-tenths of a second.

INTRODUCTION

The Importance of Respect

Humans have long dominated all other species—but as new research reveals the insights and skills of others in the animal kingdom, it's high time we also practice compassion and consideration.

It was Descartes who argued that humans have reason while animals do not. But in the 21st century, new findings on the animal mind have toppled that point of view. While scientists warn us not to anthropomorphize our furry and feathered friends, they now know that some species, including apes, dolphins and pigs, can recognize themselves in a mirror, indicating self-awareness. Chimpanzees—as well as dogs and elephants—make and use tools. Orcas pass down traditions through multiple generations. Parrots can understand hundreds of human words. Bees appear to have empathy for each other.

The explosion of insight into the animal mind continues to emerge from inspiring new research and a host of animal cognition laboratories worldwide. Much of the scientific discoveries reveal new understandings of common household companions. For example: Did you know that cats have accents and speak a special language to humans? Your goldfish actually knows you. And dogs are so attuned to their owners they display separation anxiety, grief and an emotion that experts now call love?

Together, the findings show that each species inhabits a unique niche on our planet, with their own life arc, sensations, emotions and extraordinary set of skills. In "Soul Mates"(page 100), we learn how and why our primate ancestors are committed to monogamy. In "Birdland," (page 160), the great avian intelligence pioneer Irene Pepperberg describes how she trained the grey parrot Alex, who—with a brain no larger than a shelled walnut—proved that members of his species could meaningfully talk to humans and even do math. And in "Aliens of the Deep" (page 184), we describe the uncanny intelligence of the octopus, with a brain distributed throughout its body and eight tentacled arms. Throughout the book, we show how animals raise their young, play, create art and beauty, achieve peace—and give us some lessons on how to live a more productive, satisfying existence on Earth.

The stories here demand that we revise our view of animals as less than human—and insist we see and treat them through an ethical lens. Like us, other species are compassionate and flexible, make plans and hold beliefs. Given this, we owe them good lives and the ability to thrive in the habitats where they belong. How can we shepherd the planet as a safe harbor for animals? Where that is not possible right now, how can we give them refuge? From the way we farm to how we treat our wilderness, animals deserve for us to take their lives into account. In "Walking With the Apes" (page 114), pioneering primatologist and anthropologist Jane Goodall questions: "How should we relate to beings who look into mirrors and see themselves as individuals, who mourn companions and may die of grief, who have a conscious sense of self? Don't they deserve to be treated with the same sort of consideration we accord other highly sensitive beings: ourselves?"

This book leaves no doubt: The answer is yes. —*Pamela Weintraub*

CHAPTER 1

Living Among Us

The animals sharing our homes and yards have unique personalities, specific memories and conscious minds.

Canine Love

Most dog owners will tell you that their four-legged friends adore them. It's in their dog's stance, the way they wag their tails, and that long, affectionate gaze. But is it real emotion?

New research shows having a dog as a child can make you less likely to have anxiety later in life.

Sherry Woodard lives in a mixed-species home. The three dogs and three cats that she shares her house with are family members in every way. It's a busy household, but for the most part everyone gets along. "I'm always checking in with them and they're always checking in with me," she says. "I know that they love me."

If she's in another room, they'll come in to be near her. "I do the same," she adds, "and when we're out and they're off-leash, I watch and see how often they come back and check in. They have fun playing and like to know I'm close by."

Woodard, an animal behavior consultant and certified professional dog trainer, runs Canines with Careers through the Best Friends Animal Society. The nonprofit teaches dog trainers, rescue groups, sheriff's departments and health-care professionals how to train rescue dogs for career work.

MEASURING AFFECTION

Woodard understands dog behaviors, and even though several studies have proved that our dogs love us, she says we need to understand how to interpret those signs and to know that what works for some dogs may not work for others. "It's a lot like human relationships," says Woodard. "We don't all like the same things. Some couples need more space than others. It could be the same with some dogs."

One trait that most people don't like is jumping, yet dogs do it as a way to show affection. "Since many dog owners don't like that behavior, we have to teach our dogs not to jump," she shares. "We don't punish our dogs. While we teach them, we show them that we love them. Dogs understand kindness."

CAN'T TAKE MY EYES OFF OF YOU

There are many ways to communicate with animals, but one of the most significant is making eye contact. "Gazing into one's eyes," Woodard notes, "isn't always positive. It can be threatening to a dog, unless you have a partnership. Studies have shown that when you have trust and love, our brains release oxytocin. That's true of dogs, too."

It's the same hormone that mothers release when bonding with their babies. "Our dogs know we love them by the way we treat them," she says. "Dogs have good memories and they know when we're being kind. Even if we're correcting a negative behavior like jumping, if we do it with rewards to motivate their behavior, they'll know we love them."

Dogs know when humans are treating them with sensitivity and kindness.

LOVE YOU FOREVER

Daisy, Woodard's 6-year-old Alaskan malamute mixed-breed rescue dog, talks to her. She makes wooing noises when they're playing together and Woodard responds in a singsong voice. "I'll ask her where's her nose and she'll cover it with her paw," she says. Daisy's a big dog who plays gently with the cats and other dogs. She was recently dragging the cat bed—with one of the cats on it—around the room. "Or she'll gently touch the cats with her paw. It's her way of affectionately playing with them," says Woodard.

Some dogs can't live with other dogs or cats. "Not Daisy," Woodard says. "She loves everyone. She loves children. Still, she's clingy and I like clingy. Not everyone does."

Not all dogs want to be hugged. "Some are stoic," she says. "It's not that they're without emotion. It's that they like their own space. They might want to be in the room with you, but not on top of you."

One dog she was working with didn't like hugs but nonetheless brought Woodard her toys to play with. "She loved to play," remembers Woodard. "We had a good time together. We have to remember that like us, all dogs are different, too."

A MATTER OF TRUST

"I've worked with dogs rescued from hoarders," shares Woodard. "I approach the fearful ones slowly. I respect their space. By going slow and with repetition, we can keep them safe and build their trust. After a while, they know we're trying to help them. They sense it. Dogs are so good at reading our body language. They know when we care and that we love them. They're sensitive to harsh tones and genuinely seem to love people with happy voices and smiles."

About 17% of U.S. pet owners have both a dog and cat—and yes, they do get along just fine.

Getting licked by your dog is a sure sign of affection.

A dog's happy, smiling expression communicates real love.

Signs Your Dog Adores You

Is your dog following you around because he wants to be fed—or is it something more? Here are seven ways your dog shows his devotion.

1

He Stays by Your Side
Sitting next to you, being in the same room, or snuggling are all signs of deep affection. Your dog can find a warm vent on a cold winter's day, but chooses to be close to you. That's true love.

2

She Leans in and Stares Into Your Eyes
No matter whether your dog is large or small, leaning in is akin to a body hug. If your dog watches you through the day and keeps you in her gaze, calmly and deeply staring into your eyes, it is, likewise, a sign of attachment and love.

3

He Gives You Kisses
This can be in the form of nosing, where your dog will nudge you with his nose, or he may lick your hands and face. Licking is also performed between dogs, where the licking dog has lower social status. It could also be a way for your dog to let you know that he respects you.

4

She Rolls on the Floor to Get Tummy Rubs
Rolling on the floor while exposing her tummy shows complete trust. It's the ultimate form of acceptance. If she rolls from side to side and wags her tail, she's asking for a rub.

5

He Smiles at You
A submissive smile shows that your dog is trying to please you. He's smiling when the front part of his lip below the nose shows the front incisors.

6

She Sighs
This shows contentment and relaxation, especially if your dog is stretched out next to you; she feels safe by your side.

7

He Eagerly Greets You
Your dog waits by the door and barks and jumps happily to welcome you as soon as you get home.

When your dog looks into your eyes it is an intimate act. There's nothing like basking in the canine gaze.

Your scent activates the reward center of your dog's brain.

A Veterinary Behaviorist Explains Puppy Love

Tufts University veterinarian and behaviorist Nicholas Dodman shares insights on emotions from our canine friends. He is the author of *The Dog Who Loved Too Much: Tales, Treatments and the Psychology of Dogs*.

How do you know your dog loves you?
It's in his eyes. When I'm driving and Rusty (Dodman's mixed-breed black mouth cur) is riding shotgun, we occasionally glance into each other's eyes. It's only for a few seconds, but it's a look of love and trust.

Looking longingly into someone's eyes is an intimate act. A study in the journal *Animal Cognition* showed dogs respond to our facial expressions, especially when they are focusing on our eyes. This shouldn't be confused with threatening eye contact with an unknown dog. This is about trust that's built over time.

Our eyes are expressive, and dogs are masters at reading our body language. They don't have words like we do, but they can tell when we're happy, sad or angry. This is one key to the way they understand us, but it is certainly not the only way or the whole package.

What other ways do dogs show us their love?
You might not be aware, but our dogs are surveilling every aspect of us. When we're in the same room, they often look up and glance at us. They're looking at our posture, listening to the tone of our voices, and seeing if we're looking at them or turning away.

They may sit next to us, lean into us or be as close to us as possible. If they did not like us, then they would not want to be in the same room with us.

Do they understand our words or the tone of our voices?
While language is a human trait, some dogs can learn what some words and phrases mean. A study in the journal *Science* reveals that dogs can learn the meaning of words especially when the tone is mixed with praise. For instance, when I say "Good boy" using a flat voice, with no inflection, the reaction is different than when my voice is higher and the same words are emphasized.

How do we know a wagging tail is a positive sign?
A wagging tail isn't a sure sign of love. In a nervous dog, it could be agitation. Look for a full-body wag. The tail should be midheight and wag slightly to the right. Researchers believe the direction has to do with the different hemispheres in the brain. So a dog that wags its tail toward the right is showing left-hemisphere activity, which equals a positive and relaxed response.

When first introducing a dog to a cat (or vice versa), be patient and give each their own "safe" space.

A wagging tail, being close to us and gazing into our eyes are important pieces. What we really must do is look at everything from your dog's nose to his tail. Your dog is picking up on everything—even your scent.

What role does scent play?
Your scent activates the reward center of your dog's brain. They know what you smell like and know if you've been out with other dogs. Their sense of smell is several hundred times greater than ours.

If your kitty headbutts you, don't be mad—it's a sign he likes you!

Kittens meow naturally when they feel hungry or have an unmet need.

Cat Talk

Those meows, purrs and snarls have true meaning. Here's a guide to what they're all about.

If you have a cat in your home, you no doubt already know felines are skilled at getting your attention—sometimes loudly, like when they decide you must get up before dawn because they are hungry. Other vocalizations, like purring or hissing, give obvious clues about whether they are content or upset.

But cats also have far more extensive ways of communication, both with each other and in their attempts to "talk" to humans. There's a multifaceted cat code that encompasses not only felines' vocalizations but their body language and even some "secret" messages they leave to show what territory (and which humans) belongs to them.

Mark Twain had it pegged: "You may say a cat uses good grammar," he wrote in *A Tramp Abroad*. "Well, a cat does—but you let a cat get excited once; you let a cat get to pulling fur with another cat on a shed, nights, and you'll hear grammar that will give you the lockjaw."

Learning to decipher cat talk has been a focus of cat lovers for generations, and in recent years, research has taken the effort to new heights. The findings have proved invaluable, helping pet owners care for felines and better understand what they want and feel. Decoding cat talk is key to distinguishing between physical and behavioral problems; recognizing when two felines truly are enemies; figuring out who's boss; and, bottom line: understanding feline companions more deeply than before.

Making the effort to communicate with cats can even convert folks who claim they aren't "cat people" into cat lovers, according to veterinarian and cat expert Drew Weigner, founder of The Cat Doctor, Atlanta's first feline-health specialty practice.

"I know a lot of people who professed to not like cats who ended up really liking them when they got to know them better," he says. "But if you don't understand cats, they can be difficult to relate to, even for some veterinarians. Learning to communicate with them is absolutely vital."

A feline's many different types of meows convey specific meanings, often directed at the human in the room.

DECIPHERING CAT CODE

Figuring out what felines are trying to say has interested not only cat fanciers but researchers, too. Charles Darwin, the British naturalist famed for his contributions to the science of evolution, studied cat sounds and postures and linked them to various emotional states in his 1872 book *The Expression of the Emotions in Man and Animals*.

In 1944, serious inquiry into what cats might be trying to say had a boost when Mildred Moelk published a study of cat sounds in the *American Journal of Psychology*. Her goal was transcribing what cat language sounded like with a phonetic alphabet, and then understanding the messages conveyed.

Today, researchers often divide cat sounds into the three primary groups Moelk suggested after doing her work. According to Lund University

Cats can invest their facial expressions with meaning—much like us.

Mother cats will respond with love and care when a kitten meows.

phonetician and cat language researcher Susanne Schötz, sounds produced with a closed mouth, including trilling and purring, tell us a cat seems content. Sounds made when the feline's mouth is open and then gradually closed (meowing and howling) indicate a call for attention and a need for food, water or some other material want. And the sounds a cat makes with its mouth held tensely open, including growls, hisses and snarls, indicate anger or fear.

Schötz and other modern-day cat decoders have taken Moelk's work further by asking whether nonhuman animals, including cats, have some kind of "languageness"—literal dialects across groups—and a more nuanced way of communicating very specific meanings with us and each other. Schötz, the author of *The Secret Language of Cats*, is currently studying the vocal sounds of cats in a variety of situations to see if there are differences in their phonetic patterns, depending on whether the felines are happy, hungry, annoyed and more. She's also researching whether cats react differently to how humans speak to them. "For example, we want to know if cats prefer pet-directed speech or prefer to be spoken to like human adults. We still have much to learn about how cats perceive human speech," she says.

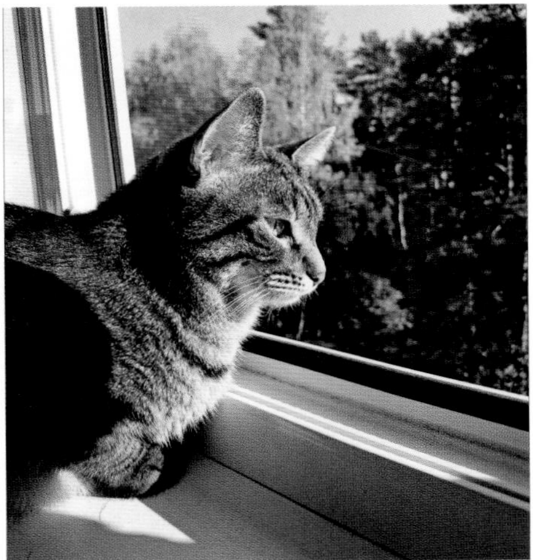

A cat with eyes wide open and ears pointed up is curious and alert, taking in the environment around her.

The meow is what Schötz describes as the preferred sound cats use to address humans. It can vary from loud and demanding (indicating hunger) to soft and coaxing (an attention for play).

Cat vocalizations typically become soft and low when frightened, and loud and demanding when they want attention—to be fed, for example. "This vocalization when they are hungry is unmistakable," Weigner adds.

THE ALL-IMPORTANT PURR

Then there is the purr, something mother cats do to calm kittens and help them sleep better. If your cat is purring near you, it's also a way your pet is showing happiness and contentment.

However, some cats purr when they are ill or in pain. You might assume if your cat is sick, it would vocalize and writhe in agony. Instead, cats' body language when they are ill or in pain is usually different. "They withdraw, get very quiet and huddle in a hunched position with their eyes closed and head down," Weigner explains. "It's more of a withdrawal than any vocalization."

If your cat begins suddenly and loudly vocalizing in the middle of the night, it's not necessarily that she has developed senility or behavior problems. Two specific, and treatable, diseases in older cats—hyperthyroidism and hypertension—are often heralded by cats loudly "talking" at night and may need a visit to the vet, Weigner says.

FACE TIME

There's more to cat talk than vocalizations. What a kitty "says" or doesn't should be viewed in context with facial expressions and body language, according to Valerie Bennett, a British animal behavior researcher and owner of The Animal Behaviour Business, which helps pet owners better communicate with their companion animals to solve behavior problems.

While humans are used to judging other people's facial expressions—like scowls and smiles—to ascertain mood, and noting body language that can indicate everything from fatigue to anger, reading expressions of furry-faced felines may seem difficult or impossible. However, using a cat facial action coding system (CatFACS), similar to one used for objectively linking human facial expressions to behaviors and emotions, Bennett and her colleagues have found they can interpret body language, vocalization and facial clues. Whether animals in their study were confined or interacting with people, signs were often the same—and frequently corresponded to whether the animals were afraid or not.

A cat with an arched back is scared and trying to look larger to make predators or other threats back off.

If you are trying to "read" a cat, pay special attention to its ears. If inquisitive, a cat's ears are usually rotated forward so they can hear better and get more information. When relaxed, a feline's ears are in a neutral position—but when scared, their ears are pulled back.

And when a cat's ears are really flattened on the side of its head, watch out: The cat is angry or seriously frustrated, especially if the back is arched and tail fuzzed out. If the cat's eyes are dilated and squinted and the feline makes a low, deep vocalization (with other cats particularly and, sometimes, at people), it doesn't mean they are going to attack—but they may.

"It's usually a frustration issue between cats that indicates 'you'd better back off or understand this is my territory,'" Weigner says. "Especially if cats are inside together, this is more bravado and posturing than a prelude to hurting each other—they are deciding who is the dominant cat."

Eyes also provide information about what a cat is feeling. "When they are wide open and looking at you, they are curious and taking in their environment. When cats are happy, their eyes are sometimes halfway closed," Weigner explains. "And when scared, they'll squint and their eyes become dilated."

CATS HAVE CHEMISTRY

Finally, cats communicate with each other through hormonelike chemicals called pheromones. Although cats, especially if not spayed or neutered, sometimes mark territory with urine,

Eyes that are halfway closed indicate contentment.

Meow Mysteries Solved! A Lexicon of Cat Lingo

This handy cat-human dictionary should give you a good start in the art of interspecies translation.

"Communication is our basis of interaction among humans and animals, too," says Atlanta cat health specialist and veterinarian Drew Weigner. "We all have a universal need to make ourselves understood. And it's how we can relate to another species." Of course, humans can't learn to literally speak cat language. It seems unlikely that meowing, howling and trilling around your pet cat is going to result in a close human-feline bond. But cats' meows and other feline lingo can be deciphered to help you understand what your kitty companion is trying to say, so that you can respond.

Here's a quick reference to help you translate cat talk:

Meow

There are several categories of meow, the most common type of cat language directed at humans for attention. The typical meow issues with a rising and falling of vocal expression in a two-syllable sound. It can be soft or demanding.

Meow Moan

You may hear deeper, darker, moanlike versions of a meow if your cat is distressed or discontented.

Meow Mew

This high-pitched meow can sound more like "me" or "we" and is often uttered by kittens or a cat who is isolated.

marking with pheromones produced by glands on their cheeks is more common and less messy (and smelly).

When your cat rubs against a chair or door—or bump his heads against *you*—he is leaving his own personal mark, via a scent you can't smell, and laying claim to that territory. He is also marking you as his human, Weigner explains—a social and affectionate gesture.

COMMUNICATING WITH YOUR KITTY

"I think there is a universal need to make yourself understood. And communication is our basis of interaction among humans and animals," Weigner says. "Most cats want to communicate. So right off the bat, as long as you are not aggressive toward them, you've solved most of your problems."

But there's a right way to approach felines, he explains. Talk softly with a high-pitched voice. "We can laugh about baby talk, but it helps," Weigner says. "And don't look them in the eye. It's considered aggressive behavior by a cat. That doesn't mean you can never look your cat in the eyes but when first meeting a feline, they may find it aggressive." Slowly reach out to a new cat; if he is already domesticated, moving quietly will help him stay calm.

Cats are no doubt smart. "They've figured it out—loud meows to make their humans feed them is behavior modification that usually works on people," concludes Weigner.

Feline Body Language Chart

How your cat stands or stares—and even the movement of his whiskers—offers crucial clues about what he is thinking, feeling and expressing.

Posture

Breathing slowly, moving in a relaxed manner—happy and content • Moving in rigid, stiff fashion or freezing—scared and about to fight or run • Moving slow, slinking low to the ground—frightened • Arched back—scared and trying to look larger • Hunched over, head down—in pain or feeling sick • Lying on back—relaxed and showing trust (could be a defensive posture, depending on the situation)

Tail

Hanging down—scared or threatened • Upright and fuzzed out—may indicate fear, trying to appear larger • Slowly moving back and forth—sizing up a situation before acting • Swishing rapidly back and forth—agitated, leave-me-alone warning • Upright and bent like a question mark—curious and likely friendly • Held vertically—friendly, likely approachable

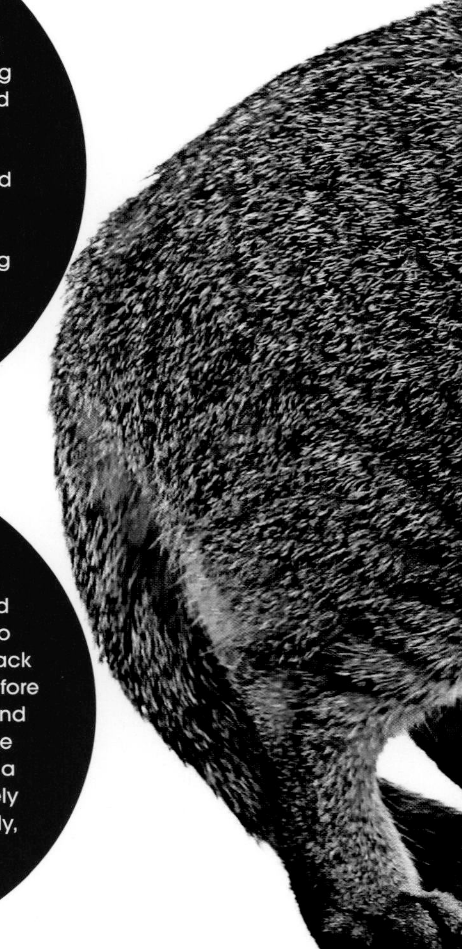

Like humans, cats have distinctive body language that speaks volumes.

Ears
Forward—content, maybe playful • Straight up—alert and listening • Turned back—irritated • Turned sideways, pointed back—nervous, anxious • Flattened back against head—angry, aggressive, scared or defensive

Eyes
Dilated pupils, eyes squinting —surprised, angry or scared • Wide open, looking around—taking in the environment • Staring—dominating, threatening or hunting • Slow blinking—relaxed, friendly • Halfway closed—happy, feeling safe

Whiskers
Relaxed, pointed to the sides—happy and calm • Pulled back, flat against face—scared or aggressive • Pointing forward—looking for prey or a toy

Cats vs. Dogs

In the contest for favorite animal, both species check boxes for smarts and emotional engagement. But which is right for you?

Whether you're a dog person, a cat person or neither, you're probably aware that arguments about which makes the better pet can become heated. Dog people will claim that canines are best because they can save lives, joke, protect their people and show true love. Feline fans will demur—depurr?—and assert that cats are smarter, have a genuine sense of humor and are self-sufficient enough to use a litter box, while dogs require constant attention. Who's right?

Whether cats or dogs are superior as pets is a matter of opinion, of course, and no one can provide evidence favoring one over the other. But some research does compare their relative intellectual capacities and ability to form attachments with a few studies claiming insight into how our furry friends might really see us. A mere comparison of these findings can't tell the full story—especially because, no matter how compelling the data, we humans tend to embrace the evidence that confirms our existing biases. Translation? Science won't change fixed opinions about whether a cat truly cares for her human family or make Aunt Lu think her poodle loves her any less.

That said, we've done an in-depth comparison on what science really says about dogs and cats when examined head to head. From smarts to emotional intuition, here's how your favorite animal companion stacks up.

THE TRAIT INTELLIGENCE

This depends on what you measure. Dogs have more neurons in the cortex ("thinking") part of the brain than cats have (530 million for dogs versus 250 million for cats; for comparison, humans have about 16 billion). Dogs are better at grasping instructions, but they also can't figure out how to untangle their legs from a leash.

THE TRAIT CARES ABOUT HUMANS

Although cats have a reputation for aloofness, they definitely show hints of caring about humans, just like dogs do. It comes down to how much time they spend with humans, especially from their early years. There's a window of opportunity for socialization when they're young and identifying with humans as part of their social circle can be solidified.

THE TRAIT COMMUNICATION

Cats adjust their vocalizations to communicate the intensity of any given need to the people who are around them, studies suggest. They can even

Both species are more sensitive to magnetic fields than we humans.

With a little help, dogs and cats can indeed be friends—but cats still need a space of their own.

dial a purr up or down to get what they want from their human—such as their supper. Dogs will, as we know, bark, whimper or whine at almost any volume in to get your attention.

THE TRAIT SOCIABILITY

Although the sociability of dogs is well-known, cats are often assumed to be antisocial. That's not the case, say researchers, who have noted that cats form social groups on their own and will "adopt" a human, even if the cat is living independently. They learn how humans communicate and obviously can be socialized to live with people and even trained.

THE TRAIT UNDERSTANDS WHAT YOU'RE SAYING

While both cats and dogs can understand something about words we repeat, dogs tend to have a bigger human vocabulary. The record holder might be a border collie genius named Chaser, who reportedly mastered the understanding of more than 1,000 human words and phrases.

THE TRAIT EMOTIONAL INTUITION

Dogs can read the emotions you are expressing and even match them to the inflection in your voice. But they might not necessarily dig into why you're expressing that emotion. So while a dog will do what she knows is "right" when you're watching her, because she can see that you have "that look," she probably won't understand exactly *why* you're looking at her. Cats don't show as high a level of emotional sensitivity, but they aren't devoid of it either. They *can* learn to read their person's cues.

THE TRAIT SENSE OF HUMOR

Both dogs and cats can be playful, given the right moment and toys. But as in people, a sense of humor may depend on the individual. Don't mistake the grin of a dog for laughter. It's more likely to be a sign of submission.

Surprising Things You May Not Know About Cats and Dogs

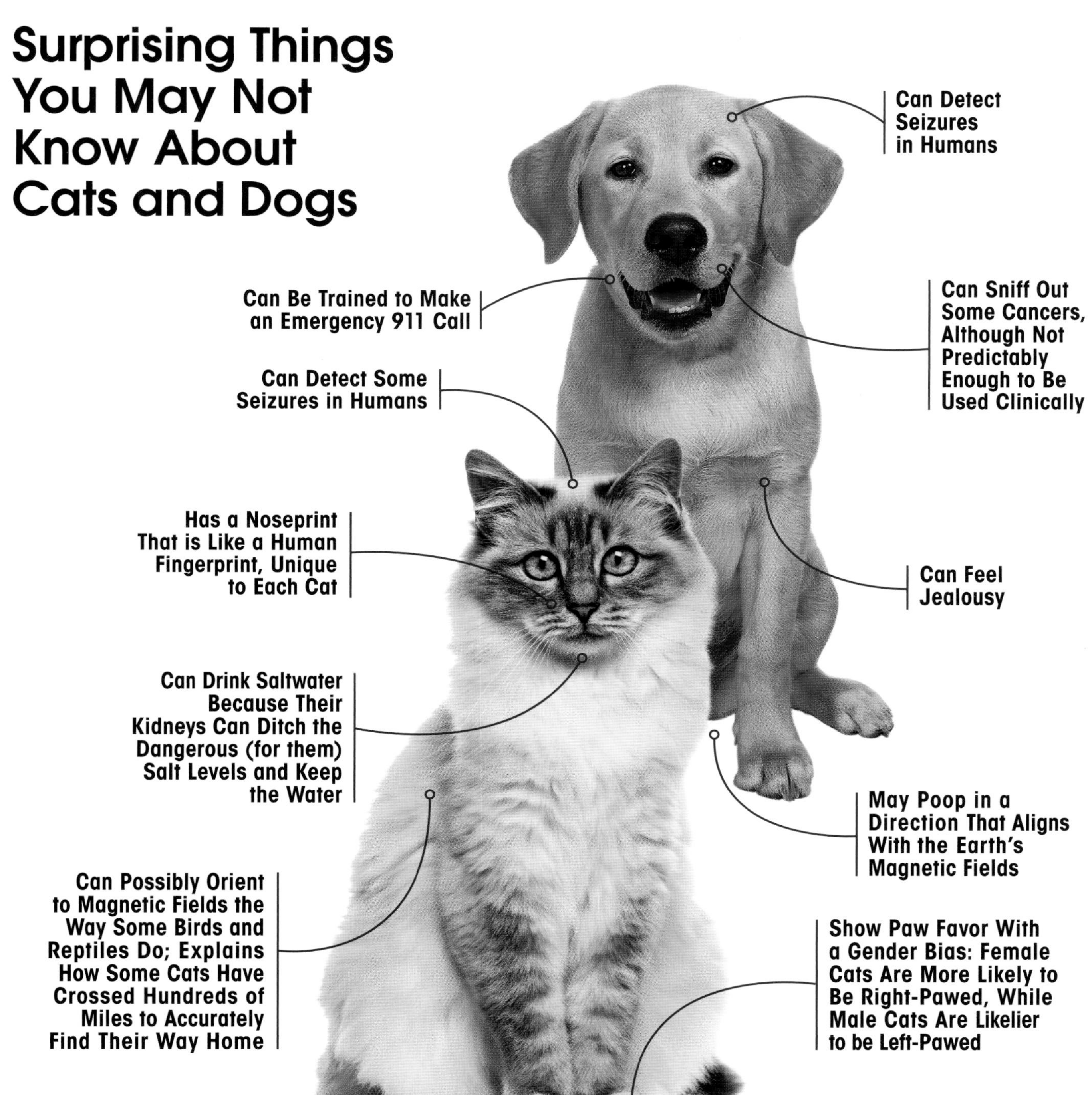

Can Detect Seizures in Humans
Can Be Trained to Make an Emergency 911 Call
Can Sniff Out Some Cancers, Although Not Predictably Enough to Be Used Clinically
Can Detect Some Seizures in Humans
Has a Noseprint That is Like a Human Fingerprint, Unique to Each Cat
Can Feel Jealousy
Can Drink Saltwater Because Their Kidneys Can Ditch the Dangerous (for them) Salt Levels and Keep the Water
May Poop in a Direction That Aligns With the Earth's Magnetic Fields
Can Possibly Orient to Magnetic Fields the Way Some Birds and Reptiles Do; Explains How Some Cats Have Crossed Hundreds of Miles to Accurately Find Their Way Home
Show Paw Favor With a Gender Bias: Female Cats Are More Likely to Be Right-Pawed, While Male Cats Are Likelier to be Left-Pawed

In triplicate: There are dog people and cat people, then there are those special few among us who love both species and have a house full of furry friends.

An Animal Behaviorist Reflects on the Nature of Our Four-Legged Friends

Erica Feuerbacher, an expert in companion animals at Virginia Tech, is lucky enough to study the behavior of animals for a living. Here, she explains how dogs see us compared to how we think they see us—with some insight into cats as well.

Do dogs view us as dogs or odd members of a social group or as their superiors or their inferiors or...? How do they view us within their social framework?

It's hard to know how they see us, but from their behavior, their relationship to us is similar to [the relationship between] infants and their parents, and they respond to us as social beings. Dogs are hypersocial, and they can likely form multiple close relationships with humans. We can't know whether they would view us as superiors or inferiors, since that would require getting into their minds. How they relate to us likely has to do with how they are reared. Dogs that are raised as family members become very attuned to us, because what we do impacts their lives greatly. Dogs that are raised more independently, such as farm dogs, likely find less use for humans.

When our dog acts excited to see us—is that for us or because we feed her, pet her and let her sleep in one of five beds she has?

I don't think these are mutually exclusive; dogs relate to us and find us important in their lives because of what we give them—the same reason that when we start dating someone we bring them things, [like] food or movie tickets.

How do dogs and cats differ in their perception of their relationship with people?

Perception is impossible to know—we have access only to their behavior, and we then put our own perceptions onto their behavior. We think of cats as more independent, but whether they really are is hard to determine because we rear cats much differently than dogs. We don't take them to kitten classes, although cat classes are starting to emerge, and we don't require they be obedient. They certainly can be, but since most people don't require this, as we do with dogs, their behavior is different. There are plenty of very attached, obedient cats, so it's possible.

When a dog licks you, what does that really mean?

We see licking in wolves when a pack member reunites, and it seems affiliative. It is likely also a way for the wolf to collect olfactory information about where the other wolf has been and what it has been up to. For dogs, it is likely similar—a greeting and a way to collect information. Other times, it is likely because they like the lotion we have on!

How can we get our dogs to do what we ask?

We need to be consistent, and that means having consistent rules. For example, if you don't want your dog to jump on you when you are dressed up, you can't let your dog jump on you when you are in your gardening clothes. It's unfair because you changed the rules. [We must also use] consistent words when we want our dogs to do something—I see owners who forget their dogs don't understand synonyms. Training your dog to come when you say "come" but then saying "over here" and wondering why your dog doesn't respond is again unfair to the dog! Use the same words all the time.

Are You a Dog Person or a Cat Person?

Our experts felt reluctant to address the question of whether dog and cat people exhibit some fundamental differences, so you'll have to take this quiz (for fun only!) to decide for yourself which one you are.

Which movie do you like the most?
A *Up*
B *Stuart Little*
C Who's got time to sit around at home and watch movies?

The best way to spend a Friday night is:
A Hanging with friends
B Reading a book, all cozy in bed
C The way I spend all my other nights: working

How do you greet friends?
A Big hugs!
B Give a slight nod, maybe say, "Hey, what's up?"
C I don't have friends

If you couldn't have a cat or dog, which of the following would you choose?
A Lizard
B Fish
C Pet rock

When looking for a mate, you prefer men or women who are:
A Affectionate and attentive
B Confident and independent
C Long distance

What do you value most in a friend?
A Loyalty and honesty
B Intelligence and an affinity for yarn
C Deep pockets

Your favorite food trend is:
A Eggs are back!
B The bug craze
C I just eat what's easy

Do you like the outdoors?
A Yes
B No
C What is "the outdoors"?

When you're working on something assigned to you, you:
A Are a rules person—you always do exactly what the instructions say
B What are rules?
C Give up

Your friends view you as:
A Social and engaged
B Quiet and often unavailable
C What friends?

Tally up your replies. If you answered mostly with A, you might be a dog person, someone who likes socializing with friends, hanging out with people, spending time outside and having a constant companion by your side. If your answers were mostly B, cats might be your companion animal of choice, to match your low-key, quiet, low-maintenance lifestyle. Did you choose mostly C? Instead of a furry pet, you might consider a goldfish.

Homeward *Bound*

Dogs and cats lost far from home rarely make it back to their humans. But there are exceptions—including these remarkably resilient, determined pets.

If your dog is lost, file a report at every shelter within a 60-mile radius of home.

A dog navigating home may be motivated by the thought of his human.

Despite the heartwarming movies you may have seen about dogs and cats making perilous treks to reunite with their humans, real-life stories of pets lost far from home often don't have those types of happy endings. But that doesn't mean they don't happen.

There are numerous documented (and many likely apocryphal) accounts of lost feline and canine companions coming home, sometimes after navigating many miles over days or months.

But how does this happen? Do they have some biological built-in GPS or compass? Are their senses so sharp they can locate home, even far away? Nobody knows.

Little research has been done on the subject. But there's speculation that dogs and cats are good at taking environmental cues to find their way home. And they may possess certain abilities—akin to other animal species that can navigate long distances—that haven't been uncovered yet.

LOADS OF QUESTIONS

There are several ways of looking at lost animals, says Wailani Sung, a veterinary behaviorist at the San Francisco SPCA. "Are the animals lost because they are taken away from the home environment and got lost in a new one? Or did they wander away from home and we presume they are lost?" she asks. "In the latter, are the animals truly lost or do they prefer to be in a different environment? Or did something bad happen to them?"

When Sung lived on a 40-acre cattle farm while she attended veterinary school, she had her own personal experience with a lost kitty. She had several indoor-outdoor cats at the time. One of them liked to wander from home and explore.

"When it was time to go to bed, if she was not home, I would call her name and within 15 to 20 minutes she would appear. But twice she did not come," Sung recalls. "The first time, she came home the next day. The second time, she did not come home for five days. How did she manage to make her way back? I don't know. All I know is that I called her every night. Did she orient to the sound of my voice? I would like to think so—but I truly don't know."

THE NOSE KNOWS

When it comes to finding their own homes and humans, dogs and cats appear to possess an innate sense of direction combined with the ability to pick up familiar smells.

"How successful an animal is in finding its way back home depends on several factors, particularly distance. Within a several-mile radius of the home, a dog (and to some extent, a cat) uses smell," says veterinarian Bonnie Beaver, a professor in Texas A&M's department of small animal clinical sciences. "Longer distance is tracked, first by north/south/east/west general orientation, until familiar smells are picked up. But how an animal can go extreme distances is unknown."

So, when a lost dog does find his way home, it may seem like the canine simply locks into the smell of the home or human he's looking for and keeps following one scent. But the canine sense of smell, combined with dog behavior, is more complex than that.

A lost dog typically starts a search for a scent it recognizes and keeps sniffing until it encounters an even more familiar odor. "Then it will orient to that odor and randomly search in the zone of that odor until it picks up a second odor that

Georgia's Amazing Hike

A mixed shar-pei dog named Georgia, 8 years old, loved to explore the great outdoors with her owner, Kris Anderson. In June of 2015, the pair were hiking in California's Los Peñasquitos Canyon Preserve. Georgia was off her leash near Anderson when she spotted some rabbits and took off chasing them. Despite many days looking for Georgia and calling the dog's name, Anderson couldn't find a trace of her beloved pet. To make matters worse, local authorities alerted her to the sad truth—the region was home to packs of coyotes and the odds were slim to none Georgia would survive for long. Then, nine days later, Anderson woke up to find someone familiar had gotten into bed with her. "I couldn't believe it," she told the *San Diego Union-Tribune*. Yep, it was Georgia, who had managed to travel 35 miles to return to her home in Carlsbad, California. The dog had pushed open a side gate and let herself into the house through the backyard doggy door. Other than some weight loss, scratches and temporary exhaustion, Georgia was fine.

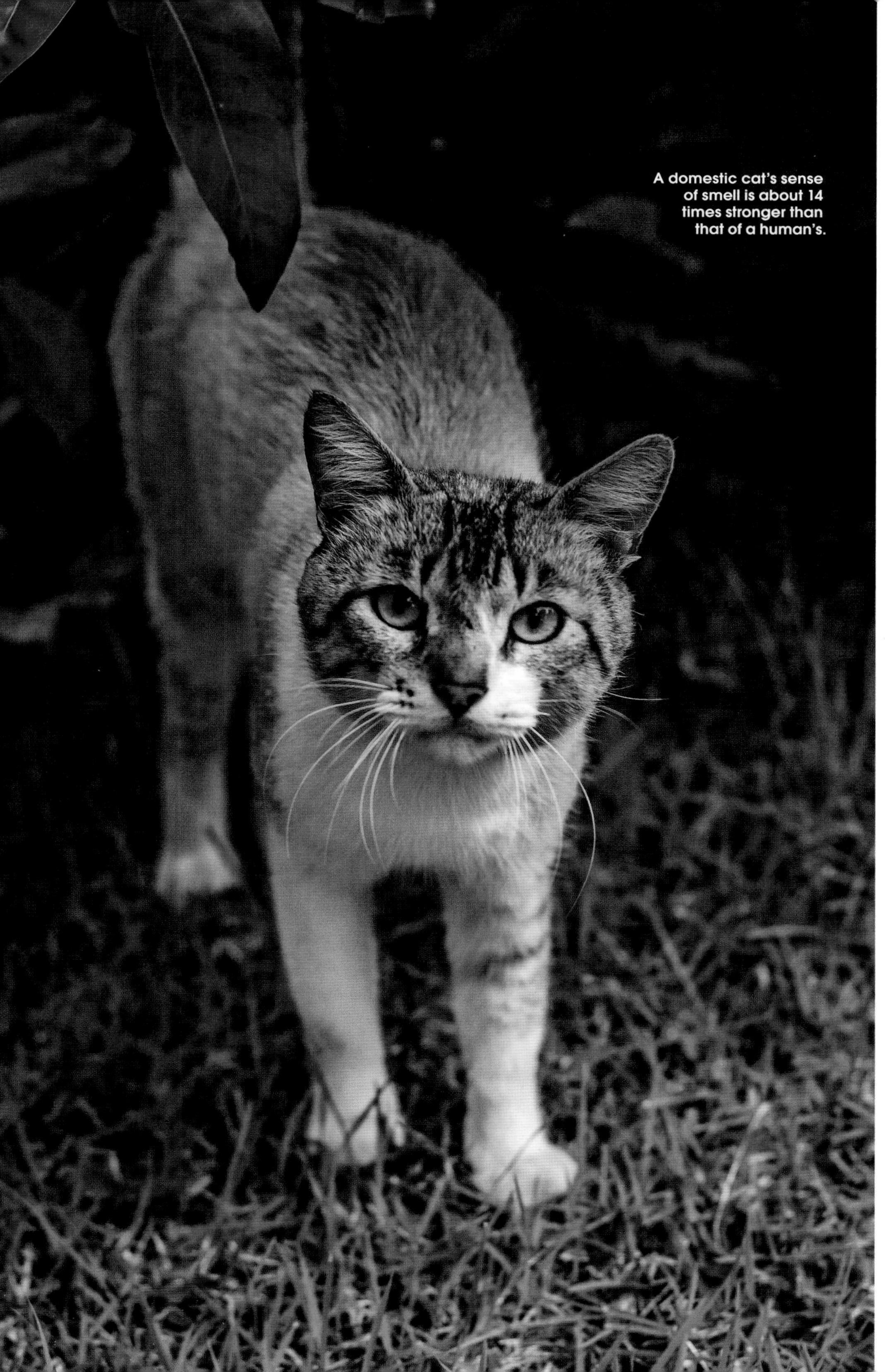
A domestic cat's sense of smell is about 14 times stronger than that of a human's.

Holly's Big Adventure

Jacob and Bonnie Richter brought their 4-year-old cat, Holly, along on a road trip to Florida's Daytona International Speedway in November of 2012. Traveling and staying in the Richters' motor home was no problem for Holly, who was used to living her life happily indoors. However, when fireworks were set off near the Richters' parked motor home one night, the cat was terrified of the noise and ran outside through an open door. Frantic, her humans made flyers, called local authorities and spread the word their fur kid was missing. After several days, the Richters had to return home to West Palm Beach, distraught their kitty seemed lost forever. But the new year brought a welcome surprise. More than two months after Holly had disappeared, an emaciated, exhausted cat who couldn't utter a single squeaky meow showed up in Barb Mazzola's Palm Beach yard, less than a mile from the Richters' home. "She was so skinny, all bones and weak and she could hardly walk," Mazzola told ABC News. The kitty was taken to the vet where a scan of an implanted microchip revealed Holly's identity and the Richters' contact information. Finally, after traveling for 62 days in the correct direction of her home, 190 miles from Daytona Beach, Holly had come back. "It was quite a journey for this little girl," Jacob Richter said. "We just can't believe she came home."

is somewhat more familiar," Beaver explains. "Next, it randomly searches the second zone until it picks up yet another, third odor that is even more familiar, eventually working into an area where all scents are familiar. And the dog then knows where to go toward home."

However, it's not only smell that dogs—and cats, too—use to figure out how to get home. "Depending on where they are lost, dogs may rely on visual landmarks to help guide their way home," Sung points out.

While there hasn't been specific research about how cats navigate long distances, Sung adds studies have determined felines use olfactory and visual cues to find their way through mazes. That indicates cats likely use both sight and smell to navigate their way back home.

ZEROING IN ON HOME

Cats, like dogs, have a keen sense of smell, but it's likely used within the relatively small area of their own territory, according to veterinarian Drew Weigner, president of the Winn Feline Foundation. But there are some reports—including one well-documented case of a cat traveling nearly 200 miles to return home along the coast of Florida—that need more explanation. (See page 35.)

"There are several types of location tools that animals or people could use in this situation, where the key factor here is the ocean," Weigner explains. "On the trip to the original destination, it's possible memory of driving along the ocean played a role. The cat's sense of direction would indicate the ocean needs to be on the other side for the return trip home."

There have also been suggestions that cats, like other animals, might use magnetic fields—a kind of "inner magnet"—to find home. The idea is based on studies showing that the ears of most mammals contain iron. "That might play a role in cueing them into the magnetic direction in the ground. There's work showing that cattle, deer and voles tend to orient in a north-south direction," Beaver points out.

Of course, nobody wants their dog or cat to go missing. But it's not uncommon: About 14% of dogs owned in the U.S. get lost at least once.

Preventing a dog or cat from getting lost involves some commonsense precautions. But microchipping and registering the microchip is the single most important thing a pet owner can do, says Beaver. "No one wants their pet to get lost and will take reasonable steps to prevent it, but accidents do happen. Thanks to microchips, lost pets can be reunited quickly with their owners."

If your dog is lost, continue to search aggressively for six weeks. Sometimes a lost dog is found after months or years.

Ginger's Snowy Trek

Some cats are lost on trips or during a move, but Ginger, an orange tabby cat, was spending time at the local veterinary clinic in Nottinghamshire, England, after being dropped off by his owner for some treatment. Escaping his cage and sneaking out a door, the cat disappeared. Although he didn't have to traverse hundreds of miles, the 5 miles to his home in Arnold, Nottingham, were anything but easy. First there was the weather, with blizzard conditions and snowdrifts blanketing much of the area. To make the cat's chances of finding his way home even more unlikely, he had to cross 30 roads and a highway. What's more, if the folklore about cats having nine lives has any truth, Ginger had already used up one or two of his lives before his mission to go home began. The tabby was navigating with a serious handicap—he only had one eye. Ten days after Ginger had made his escape from the vet's office, his owner, Jayne Middleton, noticed a motion sensor light turn on in her backyard. Checking to see what triggered the security light, Middleton made a happy discovery. Her one-eyed kitty was walking through the yard, headed to his warm house and his human."I thought I was seeing things," Middleton told her local paper, the *Mirror*. "Not only has he made his own way home, but he has done it in the hardest weather conditions."

When your dog is lost, put up flyers and don't give up the search.

LOST DOG
Friendly, white with one brown spot.
Answers to Harvey. Call 744-555-0129.
REWARD if found and returned.
744-555-0129
744-555-0129
744-555-0129
744-555-0129
744-555-0129
744-555-0129
744-555-0129
744-555-0129

Your dog can follow a familiar scent for 10 miles to find his way home.

An outdoor trek with your beloved dog is uplifting, enjoyable and a good way to get in some exercise.

Creature *Comforts*

Cats and dogs can alter our brain chemistry, busting stress, reducing depression and helping us feel loved.

If you come home after a rough day at work—or if you're going through any personal turmoil—knowing there's a nonjudgmental animal companion waiting to see you and love you unconditionally can calm you down and even help boost your spirits.

"There are countless mental health benefits of owning an animal and considerable scientific proof of this. This is particularly true if you have a strong attachment and positive connection or bond with your pet," says Nadine Kaslow, PhD, professor and vice chair of Emory University's psychiatry and behavioral sciences department. "Animals can help reduce or eliminate our feelings of anxiety or depression and help us feel less stressed."

Kaslow and her daughter, Sarah Dunn, PhD, have a lot in common, including an interest in how pets can help people psychologically. Both Kaslow and Dunn are psychologists who have each seen firsthand how cats and dogs demonstrate something akin to empathy to help their humans cope with stress and other emotional and psychological woes.

Dunn, an assistant professor in Emory University's psychiatry and behavioral sciences department, has two cats. She adopted her oldest kitty, now 17, when she was moving out on her own 17 years ago, a time she calls scary.

And, whenever Dunn has gone through rough times in her life, her elder cat has noticeably followed her around, trying her feline best to offer her support.

"As much as she wanted attention, she clearly knew that I wanted and needed her attention. She doesn't like the sound of people crying and anytime that I'm crying, she will come up to me and dab me with her paw, like she's saying 'What's up?' And it will make me smile and pet her," Dunn says.

Kaslow, who is past president of the American Psychological Association, reports many of her patients have benefited both emotionally and psychologically from adopting their pet.

"For some people, a pet is like a companion and helps them feel less lonely. For other people, a pet can help them stay calm and feel less stressed out," she explains. "Patients of mine have used their pets to help them do things they need some encouragement to do, like exercise or socialize. And for many people, taking care of a pet and spending time with their pet boosts their mood and makes them less focused on their troubles and worries."

Kaslow has also witnessed her two cats display what appears to be feline empathy for people—including patients the kitties could see only virtually.

"What I found, while being home more and doing telehealth with my patients during COVID-19, is that I've become much closer to the two cats that live with me. More than that, they're like therapy pets," says Kaslow. "When they sense that a patient of mine is distressed, for example, if they hear them crying, the cats come sit with me, and even lick the computer screen to try and comfort the person."

WHY PETS ARE STRESS-BUSTERS

When stress mounts, so can many psychological symptoms like anxiety, depression and even panic disorders. One way to measure stress

Did You Know?
Rabbits can make ideal therapy pets and emotional support animals, provided they have been vaccinated and are at least 6 months of age. Rabbits are not only adorable, but they also recognize their owners and bond quickly with humans.

There's nothing as peaceful as taking a nap alongside the family cat.

A cat that feels secure in your touch will purr with contentment and a true sense of satisfaction.

is cardiovascular reactivity, which is marked by how fast your heart beats and your blood pressure rises when you are under stress, and how quickly you recover when the stressors are reduced.

In one study, researchers at the State University of New York at Buffalo found that people with pets had significantly lower measures of reactivity when exposed to stress than those who didn't own pets. What's more, if the research subject's pet was actually in the room during the stress experiment, recovery was even faster.

One way that having a pet can directly impact your body's stress reaction is by triggering changes in hormone levels. Our breathing and stress responses calm down when we stroke a purring cat or cuddle with a dog, causing the stress-related hormone cortisol to plummet.

At the same time, interacting with pets can also trigger your brain to release the feel-good hormone oxytocin. "Oxytocin helps you feel calm and it can create a bond with your animal that can last a lifetime," notes Dunn. "So, when you are with a pet who helps you calm down, your heart rate and blood pressure go down, and you release other anti-stress chemicals in your brain."

Five Ways Pets Boost Your Physical Health

1

You'll finally have to get regular exercise. Cardiologist Michael Lloyd, MD, associate professor of medicine at Emory University, encourages patients to walk daily, noting it's a proven, health-boosting aerobic activity most everyone can do. But too many couch potatoes have difficulty committing to a daily walk. The solution? "Adopt a dog that needs walking. It's a daily requirement, which equals regular exercise," says Lloyd, whose extended family includes two rescue pups.

2

Pets help lessen the impact of chronic pain as you age. The National Poll on Healthy Aging, conducted by the University of Michigan and sponsored by the AARP, found 55% of adults 50 and older have a pet and over half have several. More than 70% of older adults said their pets help them cope with physical symptoms associated with aging. And when aches and pains of arthritis or other ills flared, about 50% noted being with their fur kids took their minds off their discomfort.

3

Adopt an overweight pooch and you can lose weight together. Researchers at Northwestern Memorial Hospital in Chicago point out there's a dual obesity problem in the U.S.—people and dogs who are seriously overweight. But when obese pet owners received nutrition and exercise counseling for themselves and their dogs, both the humans and their tubby pooches exercised more and lost more weight over the course of a year than those who were without a portly canine companion.

4

It may sound fishy, but an aquarium can help you control your blood sugar. Teens diagnosed with Type 1 diabetes must regularly check their blood sugar levels to adjust the amount of insulin they need—and, surprisingly, caring for pet fish can help. A University of Massachusetts and University of Texas Southwestern Medical Center research team found that teens with pet fish developed more discipline about regularly checking their blood glucose levels daily, which is critical for maintaining their health.

5

Pets can help lower blood pressure and cholesterol. According to the American Heart Association, multiple studies have found owning a dog or cat can help lower resting heart rates and blood pressure. In response to stress, pet owners experience smaller increases in heart rate and blood pressure. They also recover more quickly. Those who had dogs were less likely to have elevated cholesterol levels and Type 2 diabetes than non-dog owners—and were also less likely to smoke.

Canines and felines can support each other, but give them time (and space) to forge a bond.

"When we play with our dog or cat, it also can increase our serotonin and dopamine levels [neurochemicals involved with regulating mood], which help us feel calmer and more relaxed," Kaslow points out.

AN EVERYDAY ELIXIR

The chemical boost we get from pets makes them the perfect medicine for life's everyday bumps.

Even if you don't have any serious health or other problems in your life, according to a study published in the *Journal of Personality and Social Psychology*, owning a pet conferred advantages across a host of psychological areas. Specifically, pet owners had more self-esteem, were more physically fit and tended to be less lonely. They were also more conscientious, extroverted and less fearful compared to people who didn't have pets.

Caring for a pet can also provide a sense of purpose and help you feel needed and wanted. "When people feel lonely, they often wish for physical contact and touch. Therefore, touching, stroking or hugging an animal can be very therapeutic and make us feel more connected," Kaslow notes. "Many people talk to their pets, and some people even talk with them about their worries and difficulties."

Dogs and cats that have been properly introduced can get along famously and be the best of friends.

If you're feeling lonely, there's nothing like taking your dog for a walk to help you meet new friends in town.

Pets can be especially positive for seniors. Benefits for older adults include getting out of the house more and spending more time being physically active and socializing, all especially vital for anyone who has lost a loved one or is living on their own.

"Animals stand for the present, which helps older adults stay in the present and be less concerned and worried about their physical problems, less upset about their losses and less worried about their aging and the bad things that might happen in their lives," Kaslow points out. Some older adults may even find that interacting with an animal improves their memory or makes it easier for them to cope with physical pain.

THE PET PRESCRIPTION

Pets are a boon to those with developmental challenges as well. For example, there's a pet benefit for children with autism and their parents. "A pet can help a child stay calm, be less afraid and try things outside their usual routine," Kaslow explains. A dog or cat can become a best friend for a child on the spectrum, assisting with learning skills like showing affection or making friends. "A pet can teach a child with autism how to play games, take turns and share. There are many ways in which the addition of a dog in the home can reduce the stress levels of parents of children with autism, too."

While Kaslow doesn't literally "prescribe" pets, she encourages people to recognize the potential emotional, behavioral, social and physical benefits of having a pet in their lives.

"I work closely with patients in getting a pet that can help meet their emotional needs or finding ways to use their pet to provide them comfort, decrease their sense of social isolation or feelings of alienation, lift their mood and their spirits, encourage them to communicate and interact with others or motivate them to exercise," she explains. "Spending time with our animals can bring us so much pleasure and joy and help us feel loved. And every human benefits from feeling loved."

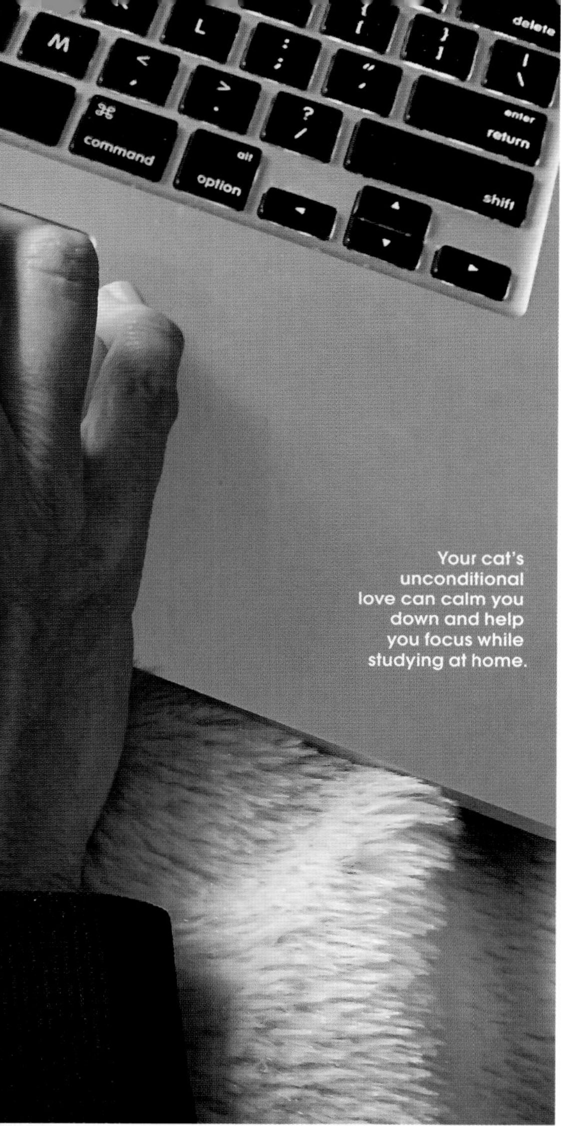

Your cat's unconditional love can calm you down and help you focus while studying at home.

Dogs especially provide so much balm because they often seem to have a sixth sense. They often just want to please you and make things better. "But unlike a human companion, they don't really expect anything in return, except your love and affection," Dunn adds. "Their love for their owner is pure. They love you and they trust you. This makes the person they love feel loved in return."

How to Select and Nurture Therapy Dogs

To help humans with stress and emotional pain, canine therapists (often confused with service dogs) must have a special calm.

Service dogs are usually large breeds highly trained to specifically help a person with disabilities; they're granted special access privileges to accompany their humans in public places. Therapy dogs, on the other hand, can be the tiniest 3.5-lb. Chihuahua, a 200-lb. mastiff or any size or breed in between. And their only "job" is doing what comes naturally to them; they must enjoy, and downright love, interacting calmly with all different kinds of people.

Requests for therapy dog visits have soared since the 1980s, and while they don't have the same public access afforded service dogs, therapy dogs are typically invited to go with their owners to visit schools, hospitals, hospices, nursing homes and other facilities where they bring furry love, attention and comfort. They may visit with elders in assisted living who miss having pets, cuddle children who have developmental problems or even give unconditional love to prisoners or those coping with recovery.

Does your fur kid have what it takes? Temperament is key. They can't fear strangers or dislike being held or petted, points out Billie Smith, who has volunteered to help people with her therapy dogs since the 1990s. "Therapy dogs can't be super excitable either. They need to be very well socialized. And no licking or jumping on people, no pulling on their leash and no aggression toward other animals or people," says Smith, who is executive director of the all-volunteer Alliance of Therapy Dogs (ATD). The dogs should also be comfortable with assisted mobility devices like wheelchairs and crutches, and be able to resist temptations like food that's fallen on the floor or that may be on a low table or lap.

"Therapy dogs have to be accepting of everyone—all people and all situations. They are 'bombproof' when they go out, not easily startled, no matter what the situation," adds Smith, who accompanies her therapy dog, a deaf Doberman whose "work" includes everything from calming down folks at airports who are anxious about flying to college students stressed from exams.

Dozens of therapy dog groups around the country are able to provide educational material about their programming, screen volunteers and dogs for therapy dog certification, and provide the necessary liability insurance for the dog and owner's therapy visits to a variety of locations. For most forms of certification, your dog needs to be at least 1 year old, clean, up to date on vaccinations, and have a friendly, attentive and calm demeanor.

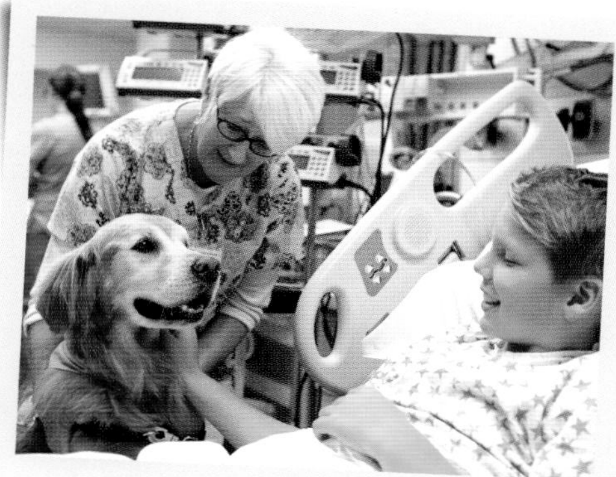

Indoor cats generally have much longer life spans (12 to 17 years) compared to outdoor cats (2 to 5 years).

Fast Tracks

Our beloved cats and dogs unfortunately age more quickly than we would like—rarely living beyond 20 years. Here's what to expect and how to cope.

Most people who have ever lived with a dog or cat—or any furry pet—know that these loved ones don't live as long as humans do. But what is really happening with our beloved pets that makes their average life spans so much shorter than our own?

The life of a dog or cat may look accelerated from our perspective, but Jerry Klein, chief veterinary officer with the American Kennel Club, says that "all species are genetically programmed for a certain period of time, whether they be dogs, horses, hamsters or humans." Most dogs, observes Klein, don't tend to live past 10 to 15 years of age (see sidebar, right, for the average life spans of our favorite pet species). Cats live an average of 14 years.

Some dogs and cats defy the odds. The Guinness World Records website gives the record for longest-lived cat to Creme Puff, who died in 2005 at 38 years old. The record holder for longest-lived dog is a bit of a controversy: Guinness lists an Australian cattle dog named Bluey, who died in 1939 at the grand (dog) age of 29 years and 5 months. But another Australian sheepdog, Maggie, died in 2016 at what her owner says was 30 years of age. Maggie didn't make Guinness because her owner lost her paperwork.

A rule of thumb that goes back to the time of Aristotle says larger animals

Each species is genetically programmed to live a given number of years, but an individual can defy the odds.

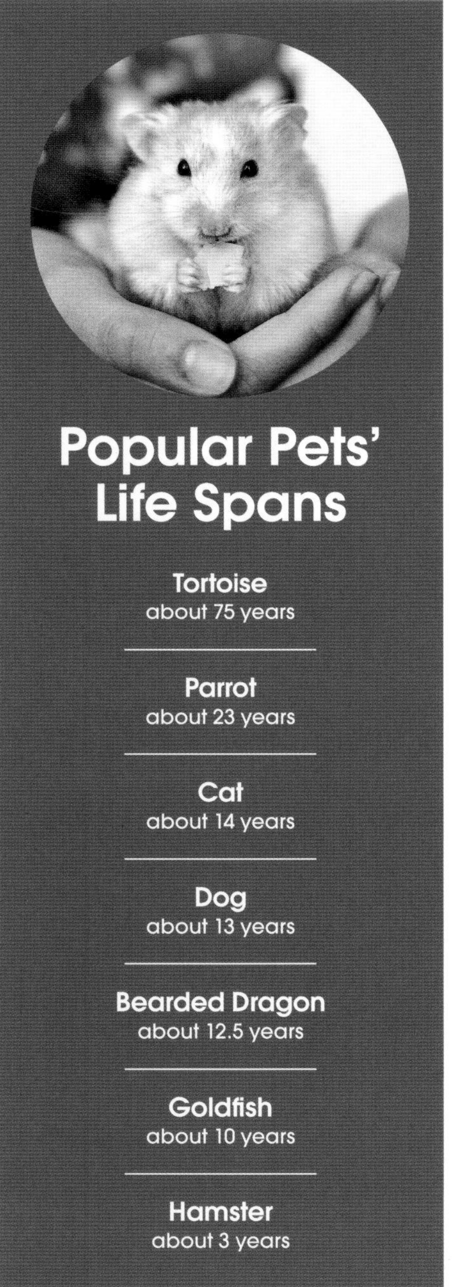

Popular Pets' Life Spans

Tortoise
about 75 years

Parrot
about 23 years

Cat
about 14 years

Dog
about 13 years

Bearded Dragon
about 12.5 years

Goldfish
about 10 years

Hamster
about 3 years

tend to live longer. But with pets, that's not necessarily the case. Cats are smaller than dogs, yet have longer life spans, in part possibly, because they prefer solitary lifestyles and thus avoid infections. Big-breed dogs, like Irish wolfhounds, typically live much shorter lives than tiny dogs, like Yorkshire terriers.

With dog breeds, rapid growth and the hormone that drives it might make larger animals more vulnerable to health problems, according to research at the University of Washington's Dog Aging Project, and large breeds do tend to be more prone to certain diseases. Dogs, whatever their breed, are all the same species—purebred dogs seem to develop health problems and to die at younger ages than mixed breeds.

DOING BETTER

Whether cat or dog and regardless of breed, our pets do seem to be living longer than in the past. The American Veterinary Medical Association (AVMA) traces these lengthening life spans to better veterinary care and improved diets. But along with those longer lives come the complications related to aging, and pet owners have to be able to navigate these issues.

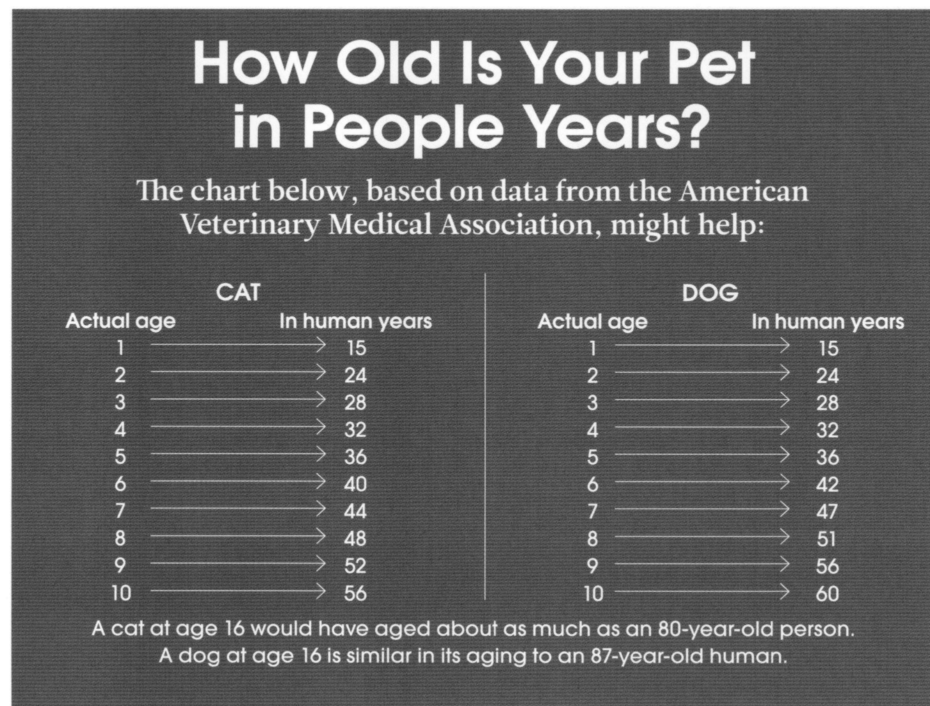

How Old Is Your Pet in People Years?

The chart below, based on data from the American Veterinary Medical Association, might help:

CAT Actual age	CAT In human years	DOG Actual age	DOG In human years
1	15	1	15
2	24	2	24
3	28	3	28
4	32	4	32
5	36	5	36
6	40	6	42
7	44	7	47
8	48	8	51
9	52	9	56
10	56	10	60

A cat at age 16 would have aged about as much as an 80-year-old person. A dog at age 16 is similar in its aging to an 87-year-old human.

As with people, dogs and cats are considered geriatric or old based on age rather than health status. For cats and small dogs, says the AVMA, the geriatric years begin around age 7, whereas larger dogs hit this threshold at about 6 years. Despite their shorter lives, our pets tend to track through the same stages of aging as humans. As they age, these animals develop problems with their teeth, eyesight and hearing. Pets can develop the same conditions humans do, including cancer; joint problems; senility; diabetes; and heart, lung, kidney or liver disease.

Pet owners often want to find ways to enhance their furry friend's longevity, but Klein nixes the possibility. "There is no magic way to stop the aging process in a dog, just as there is no way to stop aging in humans," he says. The same things that can help people stay healthy and enjoy longer lives also apply to our pets, he adds. "We can take steps to keep ourselves and our pets in the best possible health by proper weight control and regular health visits." (Note, however, record-setting Creme Puff the cat reportedly dined on bacon, heavy cream and broccoli.)

LIVES WITHOUT TRAUMA

Creme Puff seems to have had a comparatively restful life, given her indulgent diet, which might have helped. Research suggests that as with people, reducing exposure to traumas such as changing owners or family structure, or spending time in a shelter, can also help enhance life span, at least for dogs. The AVMA also

Reptiles such as bearded dragons typically live 10 to 15 years in captivity and about half that in the wild.

says that older pets might need more frequent vet visits, changes to their diets or accommodations made for them at home. Attention to weight control is also important because, as with humans, extra pounds can stress many parts of the body, including joints, kidneys and heart. Whereas, weight loss can be a red flag in cats.

As pets age, they might also start to alter their earlier habits, the AVMA says. They may be confused or disoriented, even in familiar surroundings, and show unexpected behavioral changes, such as increased aggression or anxiety. With these shifts, older pets are more likely to be taking medications and supplements—just like people.

WHEN A BELOVED PET DIES

If your pet seems to have reached a point of distress, discomfort or disease that requires consideration of euthanasia, you will face a difficult personal decision. The best person to consult when the inevitable time comes, says Klein, is your veterinarian (see sidebar, right).

Given that the life spans of our beloved pets are largely out of our control (beyond offering tender loving care), most pet owners face coping with

When Is It the Right Time?

Euthanizing a pet is agonizing. The AVMA offers a quality-of-life scale owners can use in decision-making. A score of 35 or above means the pet's quality of life is good enough to continue.

The HHHHHMM Scale covers seven criteria: Each is given a score between 0 and 10:

Hurt Is pain adequately controlled and is the pet sufficiently comfortable?

Hunger Is the pet eating enough? Does the pet need hand or tube feeding?

Hydration Is the pet sufficiently hydrated? Does it need subcutaneous fluids?

Hygiene Does the pet need brushing and cleaning, especially after elimination (pooping)?

Happiness Does the pet show joy or interest, and respond to its environment? Does the pet show fear, boredom, loneliness or anxiety?

Mobility Can the pet get up to walk without assistance? Does the pet show signs of seizures or stumbling?

More Good Days Than Bad Days When the bad days outnumber the good days, then your pet's quality of life may be poor.

The scale is a decision tool to be used with other factors and not on its own, cautions AVMA.

Genetics, nutrition and environment all play a part in how fast dogs age.

The loss of a beloved pet can cause real and significant grief—so give yourself time to mourn.

loss when their companion animal dies. Everyone manages grief differently, so the process will be highly personal.

The AVMA says after a loss, the bereaved person should not try to "rank" grief by comparing their emotions to those of a loss that might be "worse." Other coping steps include learning to adjust your self-identity as someone who no longer has this pet and seeking support from others, especially other pet owners who have experienced loss. It's also important to take the time needed to accept the reality of the loss, including facing the pain that is associated with it, says the AVMA. Finally, after the death of a pet, the bereaved person can keep the pet's memory alive through looking at old photos or keeping a journal detailing their recollections.

Sloths have such slow metabolisms, they may only poop as infrequently as once a month.

Living in Sloth Time

These low-metabolism mammals have easy, peaceful lives.

Sloths get a bad rap, but their low-tempo rhythm has kept the species around for more than 60 million years. "Their slow metabolism underlies their laid-back attitude to almost everything in life," says zoologist Becky Cliffe, who has researched sloths for a decade.

Native to Central and South America, sloths may eat eggs, insects or small vertebrates, but their primary food is leaves—abundant in the forest yet scant on caloric energy. Because their calorie intake is so low, sloths have adapted with slow metabolisms and movements.

Sloths' life spans are a mystery, but we know they spend all their time in the trees: birthing, eating, sleeping, mating, grooming their algae-covered fur, occasionally snuggling with one another for warmth, or tongue "kissing" to exchange digestive enzymes—but mostly going solo with a tight grip on the branches. A sloth that falls to the forest floor is likely to become an easy meal for a predator.

And yet this limitation can also save them. Their movements are so slow that predators often do not detect them. Sloths' languid pace forces them to be forest cartographers who must mentally map a precise path to their target leaves. Any false moves, and they could be stranded in a location without food—with no energy to escape. There's little margin for error.

But the slow metabolism also gives them peaceful lives. While males occasionally scrap over a female, sloths rarely fight; there are plenty of leaves for all. In fact, experts say, if people were more like sloths, the world would be a lot calmer place to live.

Without eyelids, your goldfish has 360-degree vision and can see everything that's going on around him.

What Your Goldfish Knows

Fish are smarter than you think.

It's easy to regard fish as little more than wet vegetables—brainless, emotionless beings dawdling in their watery world. But that view has changed dramatically in the past two decades as scientists investigate their cognitive talents. Researchers now know that fish can learn by watching other fish (an ability called social learning), recognize other fish they've spent time with, have excellent memories and personalities, figure out the solutions to maze tasks, make and use tools, discriminate among human faces, and can even distinguish between quantities (a kind of "counting").

We're likely to regard fish as brainless largely because of their expressionless faces; they seem to react to events with the same dull, uncommunicative look. As primates, we expect others' faces to give us at least a hint of what they're thinking and feeling. But fish lack the facial musculature and nerves required for those kinds of displays. Their need for streamlined bodies and faces—a requirement for living in water—outweighs any need for our kind of facial expressiveness.

Yet fish are communicating, but through signals we've only recently learned to detect and understand. Several hundred species of electric fish in South America and Africa, for example, project electric fields around their bodies, which they use to navigate and communicate. Through these signals, they tell others of their species, their sex and rank in the social hierarchy. They can let potential mates know they're attracted to them and warn rivals away; some males even engage in electric signal arguments. But fish don't need to be endowed with fancy electrical-signaling cells to communicate. Species may vibrate their swim bladders, jiggle their gill covers or even fart. Some use pheromones (chemicals that act like hormones) and visual signals to convey messages, telling others about predators or sending out "come hither" invitations to the opposite sex. Fish also have a keenly developed sense of hearing, and many use croaks, purrs and pops to utter vocal messages.

POWER OF THE FISH BRAIN

Most of us are also likely to think that fishes' small brains must be primitive, but scientists say we are mistaken. While their brains don't have a cortex—which is the key to our human language abilities—they are lateralized, a structural arrangement that enables the brain's two hemispheres to perform different tasks and to solve a variety of problems they may face.

To test fishes' problem-solving abilities, one researcher turned to the rainbowfish, an aquarium species endemic to Australia and New Guinea. He challenged the fish with a net that he moved from one end of a fish tank to the other. The net had a single hole in it, and the fish needed to find the hole to escape. He tested small and large groups of rainbowfish, and found that on average, they required a mere five attempts to pass the test. The larger groups did better than the smaller ones, a sign that fish were learning by watching each other. And the fish remembered their skill when challenged again 11 months later.

Did You Know?
The idea of keeping fish as indoor pets dates back to 1853, when aeration and water filtration systems were first designed. Today, more than 12.5 million U.S. households keep freshwater fish, making them the top pet choice! Tetras, goldfish and tiger barbs are among the most popular species.

This Mediterranean rainbow wrasse forages for food like sea urchins and gastropods on the ocean floor.

BORN NAVIGATORS

Because of the rocky streams they inhabit and long migrations they undertake, many fish have superior spacial and navigational abilities. They can accurately figure out their position in 3D—including vertical dimension, something we terrestrial animals often have trouble doing.

Fish may also have something like "place cells"—neurons that fire when the animal occupies a specific location in its habitat. These cells have been found in rats and are thought to enable the animal to form a neural map and memories of where it lives. They also likely help fish form and remember complex spatial maps of their surroundings.

TOOL USERS

Just 60 years ago, humans were considered the only toolmaking and -using species on Earth. Now we know that great apes, many monkeys, otters, dolphins, elephants and birds also make and use tools. So do fish. Brightly colored wrasses, which live among saltwater corals, use rocks to crush sea urchins to get at the meat inside. South American cichlids and some catfish glue their eggs to leaves and small rocks. If nests are threatened, they simply pick up their nurseries and swim away.

Most impressively, archerfish, which live in the mangrove swamps of Southeast Asia and Australia, spit water "bullets" to knock insects into the water. They even modulate the diameter of their mouths to adjust the shape and speed of their bullets, depending on the

Guppies watch their peers to see where prey may show up, then pass this information to future generations.

Aquarium Characters

Fish have personalities that are particular and unique. Here's a glimpse of what you're likely to find in your tank at home.

Guppies Thinking that the small, silvery freshwater fish would most likely have simple personalities, scientists first looked at how risk averse or adventurous the guppies were. But the variations scientists found in guppies were far too complicated to be described in this way. For example, after placing guppies into a new, unfamiliar environment, some of the fish attempted to hide, others tried to escape, while others explored cautiously. Furthermore, such temperamental differences were consistent over time and in different situations. Thus, when faced with a model of a heron (a predator) in laboratory tests, those that had sought to hide when placed in a new environment now also sought shelter, while those of a more exploratory bent continued to be a bit braver—although all the guppies were overall wisely more cautious.

Angelfish, one of the most beautiful of aquarium fish, are especially loved by their owners, say those in the fish pet trade. And they return that affection, learning to recognize the person who cares for them, and even accepting a morsel of food directly from fingers. Angelfish are best kept in pairs and are known to become so attached to one another that if one dies, the survivor also often stops eating and perishes.

Pacu Fish also learn to recognize their owners and, like dogs, enjoy being petted and fed by hand. They won't learn to sit up and roll over, but they will happily splash about at feeding time.

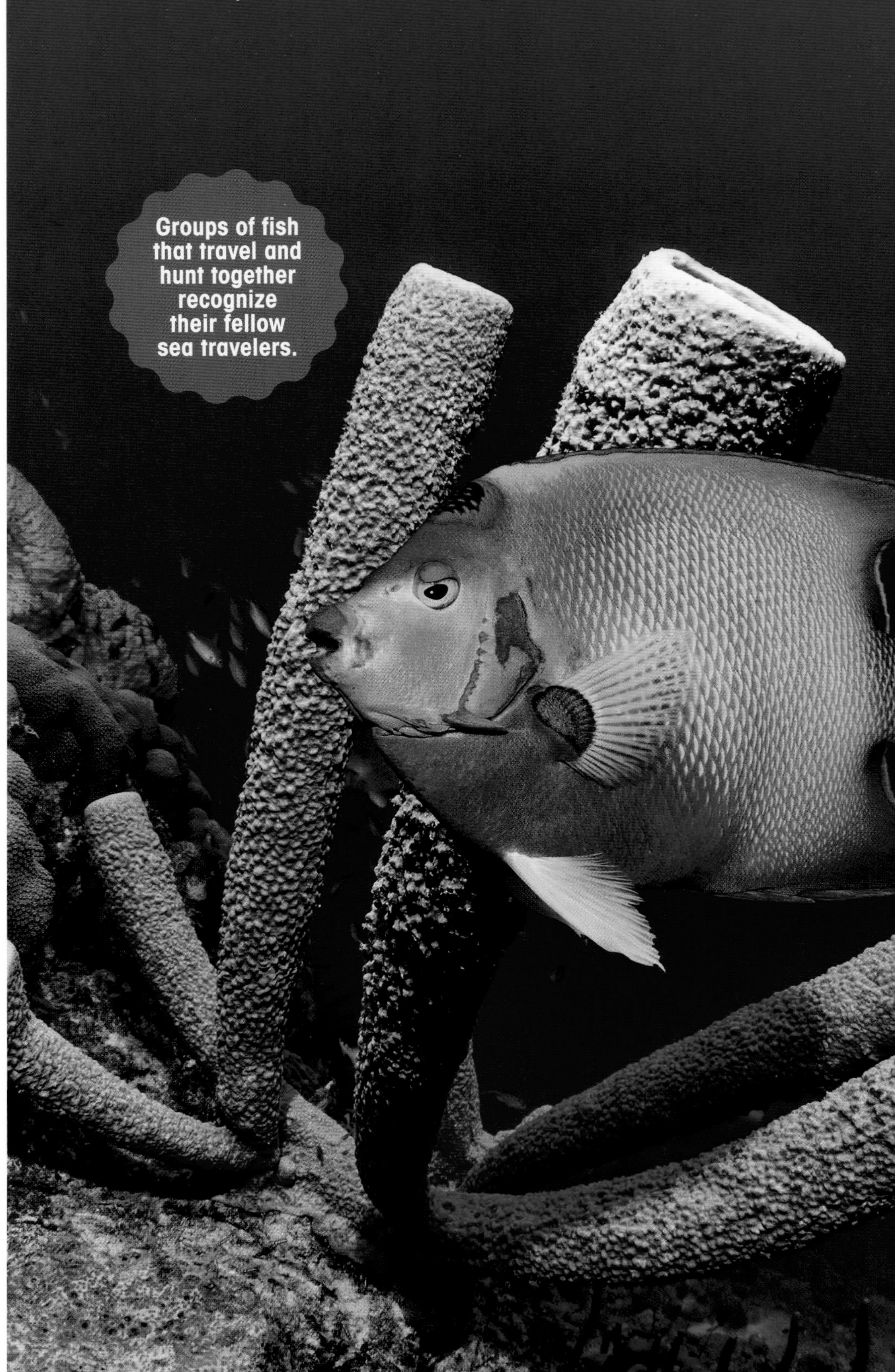

Groups of fish that travel and hunt together recognize their fellow sea travelers.

Tropical queen angelfish can grow to be up to 18 inches long.

distance of the shot—just as humans alter the shape of an arrowhead or spear point depending on what they intend to hunt.

SCHOOL FOR FISH

Juvenile archerfish learn complex hunting methods by watching experienced elders. From experiments, scientists know that most archerfish can hit a horizontal target, but they must practice to become proficient at high, vertical ones. They're also good at stationary targets, but even the best must work for days to learn how to successfully hit moving prey. They can learn the task more quickly if they first watch an expert—something scientists discovered only after removing a top shooter from the tank in the lab. The remaining four fish that had watched only up to that point began shooting—and hitting—the most difficult targets, even those that sped by like bullets. Somehow, the observing fish had been able to essentially imagine or mentally picture themselves in the position of the shooting fish. It was as if you or I had watched Michael Jordan make a tough layup shot a hundred times. Then we got up from the bench, took the ball and made the shot ourselves—not just once, but every time.

Another fish, the French grunt, hides during the day among the spikes of sea urchins. After the sun sets, the grunts migrate away from the urchins to foraging patches, following routes that are passed along culturally—as we do with our traditions. When researchers replaced the resident grunts with new transplants, the transplants didn't have a clue where to forage.

Scientists worry that many species of fish, including Atlantic cod, have lost their knowledge of where to find the best resources because we humans have removed the oldest and biggest—and most knowledgeable—members; this may be yet another reason for the collapse of cod and other fisheries.

SOCIAL RELATIONSHIPS

Because many species of fish live in schools, they're also adept at social relationships. They know each fish's position in their social hierarchy and can be as socially calculating as any chimpanzee or human, manipulating, punishing, deceiving and befriending their fellows to get what they want. Siamese fighting fish remember the male who lost the last fight and treat him like a loser. A smart male Atlantic molly that is busy courting a plump, fertile female will, if a rival shows up, turn his attention to a skinnier gal. Since males often try to woo the same lady, the first molly attempts to fool his rival into thinking that he was really most interested in the skinny female.

Mackerels hunt in groups, herding their prey. And two species of groupers hunt alongside giant moray eels, an entirely different species, rather like humans and hunting dogs. The groupers guide the eels to prey hidden in rocks and let the skinny predators flush out the fish, so both groupers and eels get a meal.

After learning that fish are intelligent, social and sentient, readers may still hunger for that plate of fish and chips. But be aware: Your goldfish does know who you are.

Pigs have so much empathy they can see things from the perspective of their human companions.

My Fabulous Pet Pig

As household companions, these supersmart creatures have been known to develop special bonds with their humans.

Does she bite?" a friend asked, seeing James Breakwell's 1-year-old playing with the family pig. "The pig, no," he replied. "The kid, yes." Breakwell, father of four, comedian and the author of *Bare Minimum Parenting*, immortalized this incident in a tweet.

Two pigs, four children and a dog make for quite a full house, especially when restrictions from COVID-19 were in place. "The kids make a lot more commotion than the pigs," he says. "The pigs are outside in the backyard most of the day when the weather is good. The kids never leave."

Pigs tend to have a bad rap—for being dirty, grumpy, rough, untrainable. However, science—and pig owners—say otherwise. The Breakwells' pigs slept in the children's bed until they were too big, hogged the blankets and allowed themselves to be manhandled by tiny hands. Both use a doggy door to go out into the yard to relieve themselves. And they use the same corner of the yard. "Pigs are more single-minded than dogs," Breakwell says. "Rather than learning tricks, pigs figure out ways to help themselves." It is not uncommon for the animals to open refrigerators, cabinets and garbage cans, and help themselves to food.

"Having enrichment is vital to keeping [pigs] happy," says Jen Reid, manager of Horse Haven and Piggy Paradise at the Best Friends Animal Society's animal refuge in southern Utah. "And by enrichment, I mean social enrichment, mental enrichment, as well as physical enrichment.... Being such smart individuals, if pigs aren't given approved ways of entertaining themselves, they'll find ways to keep themselves busy." That can mean tearing up bedding or opening cabinets and rearranging everything. "We find that things like food puzzles, outdoor time and training can help the pigs use their intelligence for good and not evil." Having an outdoor space, she adds, is essential for piggy companions to burn off some energy.

But they are also wise creatures, charismatic, gentle, loving and cleverer than human toddlers. Miranda Mastin, a

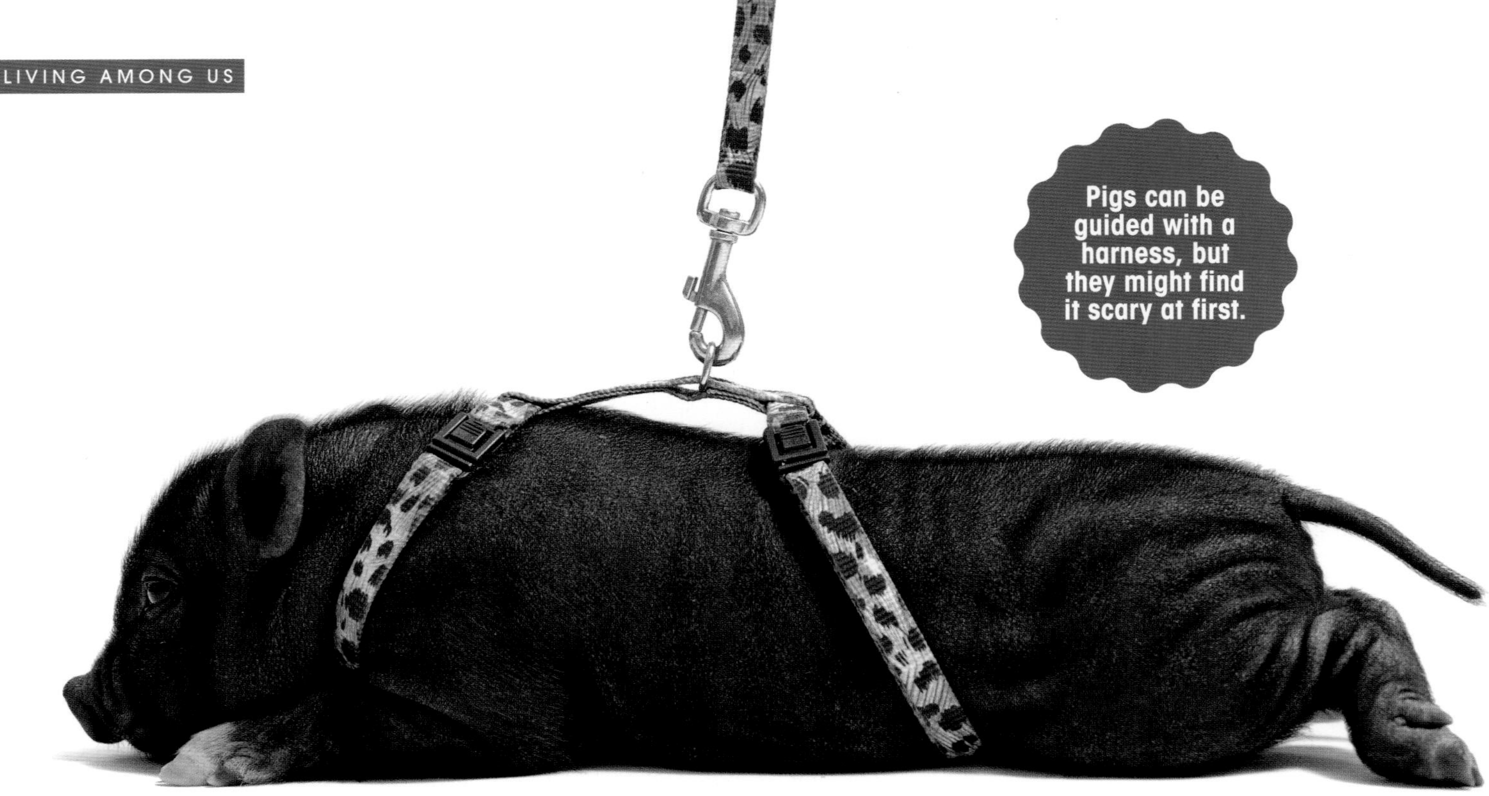

Pigs can be guided with a harness, but they might find it scary at first.

psychiatric nurse and owner of multiple pets, including pigs, says, "There's a lot of misinformation about pigs, and there's a lot of people who overbreed them. So there's an unfortunate amount of people who buy a piglet on impulse and then regret it." That's how she got one of her pigs: "I saw a post on Facebook about people giving away a 3-week-old piglet because she 'squealed too much.'"

"Pigs can be great pets for some people but not everyone," says Kali Wolfgram, a small-business owner from Minnesota. "People often see piglets and think they want one and that it will stay small. Teacup pigs are mainly false advertising, with breeders taking them away from their mothers too early and malnourishing them to try to keep them small." Wolfgram has two rescue pigs. One had been abandoned, with signs of being mistreated. The other was given away when he didn't get along with the previous family's dog. "They dig in the dirt, run around, eat grass and dandelions, and lay in the sunshine," she adds. But in winter, when it's cold outside, they have to have their own indoor activities, or they get depressed and act out.

In the wild, a few mother pigs and their piglets live together in small groups. Moms and babies sleep snout to snout.

That pigs are smart (and therefore need to be kept stimulated) is not news. Back in 2015, Lori Marino and Christina Colvin, both PhDs, wrote a paper called "Thinking Pigs," comparing factors like pigs' cognition and psychology, including their learning skills, memory, self-awareness, social cognition, time perception, novelty-seeking behaviors, emotion and personality. Their study emphasized that pigs are psychologically complex creatures with a variety of coping styles and temperaments. They are also known to show empathy.

But pigs are pigs, with "unique physical, social and behavioral needs, and if people expect them to act like dogs, the pigs and the humans will both be frustrated," says Reid. Pigs are complex—they are intelligent, messy, fastidious and loving; they need alone time, can be trained to use litter boxes and can chew holes in walls if bored.

"The one thing I would tell people who don't think pigs are pets is just to spend time with them," Mastin says. "They form relationships with people and other pigs. They mourn the death of their loved ones, they sing songs."

Did You Know?

Some researchers say that pigs are the fifth-smartest animal in the world, way ahead of dogs and cats. Pigs can play video games, perform simple math and put toys away at the end of a session. They are also capable of deception.

Not a Cat, Not a Dog

Canines and felines are not the only pets loved by humans. Beyond pigs and horses, three more enticing pet alternatives include:

Chickens

These engaging birds, which can live free-range in your backyard, are more than just producers of eggs. Research shows that backyard chickens have empathy, self-control and self-awareness. Not only do they form bonds with other chickens, they also become attached to human caregivers and may even follow a favorite human around. Of course, chickens are not house pets. They are happiest with a sky over their heads, a place to perch, a soil substrate and room to explore.

Pygmy Goats

Goats can be such wonderful companions that even Abe Lincoln kept one in the White House. Though goats respond to their owner's voices, it's rare to bring a dairy goat inside. Instead, a recent trend has been the compact, companionable Pygmy goat, which is similar in size to a large dog but has special needs. It eats grass, brush and leaves and requires a shelter of at least 80 square feet plus land to roam. And unlike a dog, who may prefer to have the house to himself, a pet Pygmy goat will thrive best with a Pygmy friend.

Rats

Many people may be surprised to learn that rats are intelligent, affectionate and, despite their reputation, keep clean. You can teach your pet rat tricks like spinning or overcoming obstacles in its path. Like your dog, a rat will seek out human companionship and even initiate play. Despite all this, you can't just allow your rat to run freely around your home like a cat. Instead, you'll need to house your pet rat in a large wire cage that is at least 2 square feet—and preferably one that is twice that size. Find a habitat with horizontal bars so your buddy can have fun climbing the sides. And no, a big glass aquarium will not fit the bill. When contained, your rat needs a cardboard nest box and clean water. A low-fat, pelleted diet made for rats should provide basic nutrition, and you can supplement with fruits and veggies on your own. Some rats are "adopted" from science labs, including those that research cancer, so when you select your pet rat, if possible, try to check the provenance to be sure he or she comes from a line that is relatively cancer-free.

Recent research finds that chameleons change colors to communicate and control their temperature.

The Inner Lives of Reptiles

We've given these cold-blooded creatures the cold shoulder for decades, but research reveals a complex cognitive landscape and rich emotional experience.

From crocodiles and geckos to snakes and turtles, reptiles have long been dismissed as coldhearted, sluggish, unthinking and unfeeling. Nothing could be further from the truth. In fact, over the course of the past two decades, important new research has established reptiles as sentient, emotional animals long misunderstood. Their new status comes from the recent ascent of reptiles as pets of choice, with ownership in the tens of millions today.

A spotlight on the state of the science comes from Helen Lambert, PhD, an animal welfare consultant in the U.K., and her colleagues. Writing in the journal *Animals*, she defines sentience as "the capacity of an animal to feel and experience both positive and negative emotions and states." Her team found 37 studies in all indicating reptiles are capable of "anxiety, stress, distress, excitement, fear, frustration, pain and suffering" and four studies showing they could also experience pleasure (among other emotions).

These studies, in aggregate, have enormous implications for how pet reptiles are treated in our homes. When people underestimate reptile intelligence

Ball pythons are perfect pet snakes because they are easy to handle and tame.

and needs, the animals can suffer considerably in captivity. "An animal can only have good or even adequate welfare if negative emotions and experiences are minimized and positive states are promoted," Lambert said.

HIGH ANXIETY AND SOCIAL STRESS

To recognize and then minimize negative emotions and experiences in reptiles, closely observe how they behave, Lambert's team notes. For example, one study showed that delays in feeding reliably appeared to cause pain in the ball python. Another study concluded that handling green iguanas precipitated an increased heart rate, an indicator of emotional stress.

Other experts couldn't agree more. Researcher Gordon Burghardt, PhD, of the University of Tennessee, Knoxville, for instance, used to keep iguanas in his lab. "At a certain age some would cower in the corners of their enclosure or would eat less because they were being dominated by other individuals. Those animals were stressed." The solution? Separate the animals.

Lack of warmth, a sense of threat and inadequate food are not just physical deprivations, they also cause emotional pain. Scientists have identified reptiles' turmoil by measuring levels of the stress hormone cortisol in the animals' feces and blood. The more anxious the animal feels, the more its cortisol levels rise.

"A reptile that's handled too much or in a substandard environment [the wrong temperature, incorrect spatial needs, facing constant threat or competition from other pet animals] will be stressed out," Burghardt says.

There are some common signs of reptile angst. "Turtles withdrawing into their shells, color changes in lizards, excess locomotion such as pacing and escape attempts, defensive posturing and decreased feeding are a few indicators that the animal may be stressed," Burghardt says. "When we get sick, our temperature rises to fight off the infection. Reptiles, being cold-blooded, can't raise their body temperature, so they seek out warmer areas, an additional reason to provide a temperature gradient."

Lizards and tortoises appear to like some people more than others. They also show pleasure when stroked.

CAUTIONARY NOTE

Pet owners, beware: Lambert and colleagues point out that "the average person handling a lizard or a turtle may be unaware of the emotional stress they are causing them." That's a cautionary note in light of increased numbers of wild-caught animals being sold in the

Bringing a Reptile Home

You're searching for a scaled or shelled companion, but which is the right pet for you?

Back in the 1950s, dime stores sold baby red-eared slider turtles and painted them as if they were Easter eggs. Usually the food the pet store dispensed was inadequate, and as the turtle grew, its shell became deformed and it quickly died.

Happily, things have improved significantly since then when it comes to the care and feeding of reptiles. "Pet stores now have pretty appropriate stuff," Gordon Burghardt, PhD, says. However, "it's essential to learn your specific reptile's needs, diet and water requirements." Also provide them with appropriately sized housing (with places to hide), and light and heat sources.

Special concerns include knowing about common health problems. Is your reptile arboreal (tree-dwelling) and in need of branches to climb, or is he ground-dwelling and in need of a burrow in which to hide?

Many reptiles will require an external heat source to establish a temperature range within its enclosure; this allows it to warm up to help digest food and calm itself if it feels threatened (by us or another of its own kind).

"We know now that reptiles need a more enriched environment—that includes places to hide, enrichment items it can explore, a variation in their diet and even problems to solve. And many reptiles are more social than we have appreciated in the past," Burghardt concludes.

The tortoise differs from other turtles because it lives on land.

pet trade, along with increased sales of captive-bred animals. Both result in pain, suffering and death for their cold-blooded stock. The estimated mortality rates for wild-caught reptiles in the pet trade is from 5% to 100%, and it is between 5% and 25% for captive-bred species, according to Lambert and team.

Even a 1% rate of death during transport in the pet trade is likely to involve millions of animals, given the scale of the industry, says U.K. biologist and medical scientist Clifford Warwick, lead author of *Health and Welfare of Captive Reptiles*.

REPTILES HAVING FUN

Of course, it's not all about avoiding pain—it turns out that reptiles can play. "This is something we think we've really documented" across the reptilian species, Burghardt says. For instance, Komodo dragons, the world's largest lizards, engage in interactions with objects such as buckets, boxes and old shoes. "Sped up a little bit on video," he notes, "their behavior is similar to that of dogs." In fact, in YouTube videos, you can see the dragons playing tug-of-war with dogs, and a tortoise and dog competing over a soccer ball. (Spoiler: The turtle wins.)

"Once researchers started testing reptiles in the right way, they found that reptiles were as good as birds and rats and whatever other animal was put through the same tests as long as the tests were calibrated for the animals' natural behavior," adds Warwick.

Lambert and her colleagues undertook their wide-ranging review in part to see what future research is needed to help maximize reptiles' captive welfare. With people beginning to realize that there's much more to reptile capabilities than we ever thought possible, that would be a good thing—for us and especially for our reptile pets.

Five of Today's Most Popular Reptile Pets

1
Red-Eared Sliders
(Trachemys scripta elegans)

Although these water turtles are native to the southeastern U.S., they've become an invasive species worldwide. Males can grow from 6 inches to 8 inches in length; females from 8 inches to 10 inches in length. With proper care, these animals can live 35 years in captivity. These opportunistic omnivores can be fed "earthworms, small fish and commercial turtle food," notes Gordon Burghardt, PhD. "Provide these active turtles with various enrichment objects. They may or may not interact with these directly, but they may climb around an object, investigate it if it has different colors and textures." He emphasizes that this species should never be kept outside of its native range in North America.

2
Corn Snakes
(Elaphe guttata)

Corn snakes (also called rat snakes) are found in the southeastern U.S. as far north as New Jersey. They're carnivorous constrictors that can live 20-plus years in captivity. They prefer to burrow and need appropriate matter to do so. "I'm a fan of feeding snakes live prey—rats or mice," Burghardt says. "It's part of their enrichment, their psychological well-being. But you must be careful; the mice or rats can actually hurt the snake if the snake is not hungry." These snakes can also be fed "thawed frozen rodents or humanely killed prey that can be moved with forceps to stimulate grabbing," he adds. This species is docile and accepts gentle handling. It will rapidly shake its tail when threatened.

3
Bearded Dragons
(Pogona vitticeps)

These semi-arboreal lizards are native to Australia and eat insects as well as plants. They're active during the day and can live up to six years in captivity. "Lots of people keep bearded dragons and they love them," Burghardt says. "Beardies need social stimulation, which they can get from people, but they can interact with dogs and cats—as long as the animals won't hurt the lizards." When a beardie feels threatened, it will flare out its beard, which darkens. This display is designed to ward off imminent threats. "You may see head-bobbing, used to communicate with other animals and sometimes in competitive or sexual interactions with a bearded dragon tankmate," he says.

4
Chameleons
(Chamaeleonidae)

Native to Africa, Madagascar, the Middle East, Europe and Asia, this colorful, carnivorous lizard does best on a diet of insects. It shoots out its tongue, which is twice the length of its body, to catch prey. Because they are territorial, solitary and don't like to be handled, they're not a good pet for beginners and should be kept alone in an enclosure. Understanding your chameleon's normal coloration is important in interpreting its well-being. Males may live up to seven years; females have shorter life spans."Only more common, captive-bred species should be purchased," Burghardt says. They need lots of ventilation and do poorly in most glass or plastic caging.

5
Leopard Geckos
(Eublepharis macularius)

These docile, ground-dwelling lizards native to Pakistan, India and Afghanistan eat a diet of insects. They're nocturnal, so if they're inactive during the day, don't become worried unless they show signs of illness. They can live over 20 years with the proper care. You might see your gecko climbing its tank. This may indicate it's bored and lacks objects to climb and interact with. Or its tank may be too small. It may also be trying to get away from a tankmate (in which case, separate them).

Zip forms a special understanding with his riders.

Zipped in Black Magic

How a blind therapy horse helps people with special needs recover their lives.

It's late fall in Pennsylvania. The trees have shed their leaves, and the horses have grown thick coats that will keep them warm through the winter. It's barely 5 o'clock, but already, twilight has stretched across the woods and light spills from the stone barn, welcoming those arriving for evening lessons with the 15 horses at this special center for equine therapy. And in the middle of it all is Zipped in Black Magic, a tall, handsome horse with the telltale dappled coat of an Appaloosa. He paces his stall, keeping an ear flicked toward the feed room, and when his favorite volunteer slides open his stall door, he drops his head to her arms, first for a hug, then to sniff what she carries. Zip, as he's called, steps sideways as she dumps grain into his feeder, then allows a last snuggle before plunging his nose into his meal.

Zip is the star of All Riders Up, a PATH (Professional Association for Therapeutic Horsemanship) accredited program for children and adults with special needs. Born in central Pennsylvania as a Western show horse, Zip's career ended early after a bout of uveitis left him blind. Sensing his potential worth as a therapy horse, Joelle Manners and Billie Palumbo, Zip's former owners, contacted All Riders Up in 2015 to see if they could donate him to the program.

"We don't want people to judge our students on the basis of their disabilities, so why should we allow Zip's blindness to cloud our consideration of him?" Marcy Laver, co-founder of All Riders Up, remembers thinking. "We knew we had a lot to learn about caring for a blind horse, but we were so impressed with Zip's demeanor."

A therapy horse must be calm and unflappable when clients become upset, frustrated or confused.

The primary quality for a therapy horse is unflappability. Therapy horses undergo regular training during which they're exposed to various situations, equipment and techniques they might encounter in sessions. For someone with muscular dystrophy, for example, Zip's gait and body temperature might be crucial in helping that person's muscles relax. The early part of each session might focus on coaxing the muscles into relaxing while engaging in stretching or strengthening work. For a student recovering from a brain injury or stroke, Zip's four-beat walk, which mimics the hip movement used by humans in walking, can help the rider regain the muscle memory and coordination needed to walk again. For a veteran suffering from PTSD, sessions might incorporate human-horse communication while helping the rider tune in to particular emotional states. Equine therapy can help some people learn to plan and make decisions; it can help others develop life skills.

Across these goals, Zip caught on and proved deft. He flew through his training and took everything in stride. He was soon trotting the pasture with Dibbs, a fellow Appaloosa who fast became his best friend, demonstrating to everyone he had found a new role and new home.

RECOGNIZING THE GIFT

Horses are sensitive creatures and can easily pick up on sensory stimuli resulting from changes in human biochemistry, vocal tones or body language. In this way, they provide critical feedback to instructors and therapists on a student's needs and emotional state and can be ideal therapy partners. For example, if a student comes to the program needing to develop executive functioning skills, an instructor will design lessons that require that student to plan and carry out a series of step-by-step instructions involving the horse. If the rider rushes the exercise or neglects to give clear directions, the horse will likely respond with confusion. The student quickly learns that each component of the exercise must be properly executed if she is to succeed. Similarly, a horse can sense a rider's fear or nervousness and will know to approach that rider quietly and calm the person down.

"With Zip and all good therapy horses, the most important quality is trust," Laver explains. "They unconditionally trust their handler and thereby become trustworthy" as partners in the therapy process.

Horses chosen for therapy work become familiar with adaptive riding equipment, which can include ramps, electric lifts and leg braces, as well as riders whose vocal or body control might be erratic. They grow accustomed to tools and activities designed for the therapeutic setting and often develop an extra sensitivity toward riders' needs.

When Zip came to All Riders Up, he learned to accept electronic wheelchairs, bouncy balls, Hula-Hoops, and the presence of leaders and side-walkers during riding sessions. He understands that a student might shout or grow fearful or move erratically while astride him—and he remains calm, regardless of what's happening. Even more remarkable, he often intuits a rider's needs before his human partners do. He lends confidence to nervous riders (and their parents). He offers students who might not succeed in traditional academic or social settings the chance to thrive. And he lends mobility to those unable to ambulate independently.

Perhaps because he is blind, Zip is especially attuned to the environment and the people around him.

Perhaps it's his own lack of sight or just his spirit, but Zip is always attuned to the environment and people surrounding him. Christina, one of All Riders Up's longest-term students, loves challenging herself by riding without using her hands or by tackling a particularly difficult trail. She might have a moment of panic, daunted by the task ahead of her, but Zip carries her faithfully on each and every ride.

Anthony, another long-term student, enjoys walks on the sensory trail, an outdoor activity set up to engage students' senses—from sight to touch to sound and more. Anthony likes to visit the music station and ring various wind chimes, bells, cymbals and rhythm sticks. Another horse might be startled by the cacophony, but Zip will level his

The best therapy horses trust humans and are trustworthy in turn.

head calmly at the instructor as though to say, "Are you OK with this?" Then he'll relax and stand patiently while the symphony plays around him.

A VERSATILE THERAPIST

So long as he has his trusty pasture mate nearby, Zip is happy to spend sun-soaked days eating grass and cantering from one end of the paddock to the other. He carries riders through obstacle courses and wooded trails, stands quietly while less-mobile students mount using a ramp and/or hydraulic lift and obliges riders developing gross motor coordination through grooming and bathing exercises. His easygoing nature puts new volunteers and students at ease, and his courage at overcoming his own physical disability inspires riders to test their own abilities. He is a reliable mount for safety demonstrations and volunteer trainings, and he's often the first horse newcomers meet. The prospect of working with 1,000-pound animals can be daunting, but an introduction to Zip often breaks down that first barrier toward therapeutic success. Until they're told, most people don't realize Zip is blind, which, as one program parent says, "shines a brand-new light over him."

One of Zip's most loyal supporters is Cooper, who was born with cerebral palsy and came to All Riders Up just shy of his fourth birthday. At first, he was unable to hold himself upright, walk or ride for more than 15 minutes at a time. A year and a half later, Cooper walks to and from the lesson arena with the aid of a gait trainer, holds himself astride and signals his horse to start, stop and turn. Cooper has increased mobility and hand-eye coordination, is able to follow multiple-step directions and is reinforcing what he learns with his teacher and speech pathologist. "You can't help but notice this horse," Cooper's mom says. "Even without being able to see, Zip still trusts. It's amazing to see Cooper bond with a horse."

CHAPTER 2

Animal Society

Many creatures live in dense groups and—like humans—love their children, have friends and mourn their dead.

Orcas are especially prevalent in the waters off the Pacific Northwest and Norway's northern coast.

Orcas are big dolphins, but ancient sailors named them killer whales.

The Culture of Killer Whales

With enduring family relations, unique dialects and diets, and huge celebrations for special life events, the orca has evolved over time, passing tradition from one generation to the next.

Jennie travels about 75 miles a day with her infant son, her mother, her teenage daughter, a grown son, sister and her 80-year-old grandmother in search of food and to explore, play and socialize. Because of the strong cultural traditions she was born into, she eats a very specific diet, communicates in a dialect unique to her group and socializes mainly within her family. Her daily activities reflect ancient cultural traditions passed down from one generation to the next. Her infant son, just like her adult son, will stay by her side for the rest of his life. Her daughter and sister will help her in raising her little one through a long childhood. And she will depend upon the experience, accumulated knowledge and wisdom of her grandmother—the matriarch—to make the right decisions when life becomes hard.

Orcas of Puget Sound are besieged by the constant roar of industrial noise disrupting echolocation and navigation.

Jennie sounds like a person with an extended family group whose members each have a valued role in a society driven by long-standing cultural traditions shaped over generations. This could be a description of you or me. But Jennie just happens to be from another species—the orca, better known as a killer whale.

Cetaceans (the mammalian order to which orcas belong) evolved from a land-dwelling animal about 50 million years ago; fossils show they gradually adapted to a fully aquatic lifestyle over 10 million to 15 million years. Orcas are actually the largest species of dolphin, a member of the cetacean family *Delphinidae* within the cetacean suborder *Odontoceti* (toothed whales). They can grow to 25 feet long and weigh from 4 to 6 tons. After a gestation period of 18 months, an infant is born that will take more than 15 years to mature to reproductive age. They are, in their natural habitat, very long-lived, with females sometimes reaching the age of 80 to 90 years or more and males 60 to 70 years of age.

Did You Know?
The standard dolphin, above, is quite small next to the orca. Both are speed demons, with dolphins clocking 33 miles an hour versus 31 miles per hour for the orca. Dolphins live for 40 to 50 years compared to orcas living 70 to 80 years.

Young orca females can help out by babysitting for the pod before making families of their own.

Orcas can hold their breath underwater for up to 15 minutes.

A pod of orcas swim through their ocean home.

Jennie, a composite female I've created based on the dozens of killer whales studied over the years, lives in the Pacific Northwest, in a highly complex hierarchical society consisting of different levels of associations and relations—all based on learned cultural traditions. Her community (known as the "southern resident killer whales," or SRKW to scientists) is actually a large extended family, known as a clan, comprising three pods named J (with 22 orcas), K (17 orcas) and L (34 orcas). Each pod, in turn, is composed of several matrilines, subgroups led by older female matriarchs. Individuals within a matriline, such as Jennie, her mother, grandmother, sister, daughter and sons are connected by maternal descent. Jennie stays mainly within her matrilineal group, but at times she socializes with other pods.

A TRAUMATIZING PAST

There are 72 members of the SRKW population living free today, with one in captivity. The clan members have a disturbing history because, from 1965 to 1975, they were routinely culled from Penn Cove, Washington, and taken for use at exhibits at marine parks. Over those years, a third of them were captured or killed, leaving the group decimated as they currently deal with challenges such as low levels of their favorite prey, toxic contaminants in the waters and excess vessel traffic and noise. They remain endangered to this day. Fortunately, most other orca clans are doing better than the SRKW and, when left alone in the wild, manage to thrive.

J35, a female member of J pod, is seen on the 17th and final day that she carried her deceased calf in the summer of 2018.

Even amid their challenges, about once a year, all three SRKW pods gather in one location in what's known as a superpod. The occasion varies. It may be a greeting ceremony or a pure celebration of life as the pods mingle, socialize and play. At other times it may more closely resemble a grief ritual after the loss of one of their own. (In 2018, shortly after a young orca from J pod went missing after a long illness and was presumed dead, all three pods came together in what some have interpreted as a group display of togetherness after the loss.)

Jennie's relationship with her grandmother is vitally important to her survival because, as the matriarch, her grandmother holds the accumulated cultural knowledge of many generations and now, as a female orca "of a certain age," is entirely devoted to helping other members of her family navigate life successfully. You see, orcas are one of the very few mammals who experience menopause. When female orcas pass through middle age they stop reproducing to concentrate more on helping their children and their grandchildren. This is called the "grandmother effect" and, thus far, it appears only in orcas, beluga whales, short-finned pilot whales, narwhals, sperm whales and, of course, humans.

SRKW usually spend their summer in the waters around Washington state and southwest British Columbia. In the winter, they expand their range to find more food because these whales are gastronomic specialists of the highest order, with chinook salmon making up 80% of their all-fish diet. This highly specified diet is a cultural tradition, too. And despite the fact that there are other kinds of salmon, other fish and mammals available, these whales are so culturally conservative that, even when chinook salmon are scarce, they will not switch to a more abundant prey. It is as if these orcas see themselves as "the chinook salmon eaters" and

Cultural differences among separate orca communities stop members who meet up from mingling and mating.

their cultural identity depends upon maintaining this dietary habit. Other orca cultures around the world possess their own dietary specializations. One community off the coast of New Zealand specializes in manta rays and stingrays, having learned techniques to grab their prey without being stung by their tails.

SONG OF THE ORCA

Individual orcas express their identity through cultural habits, including dialects—sets of calls learned from parents. Each pod uses a distinctive dialect to communicate its social identity. Some aspects of dialects are shared across pods. The more related the pods, the more similar the dialects. In short, the similarities and differences among dialects are a kind of sound map of interpod relationships.

The sound repertoire of orcas is as complex as their social relationships. They make different kinds of sounds—clicks, whistles, pulses—in various ways and contexts. Like other dolphins, orcas navigate through echolocation (sonar), producing high-frequency clicks underwater and processing the echoes that bounce off objects at frequencies from 0.2 to 150 kHz, an order of magnitude faster than human sound processing. Whistles and pulsed sounds combine with echolocation to create a highly complex acoustic landscape that only orcas fully understand.

Like their diet, the dialect of the SRKW community is unique from other orcas. The cultural differences help them maintain genetic distinctiveness as they choose to mate only within their group.

Indeed, when SRKW members meet up with transient orcas wandering into the very same waters of Puget Sound, they share overlapping territories with noncompeting diets. The transient whales travel in smaller pods than the SRKW, range wider and eat only mammals (seals, sea lions, small porpoises and an occasional otter) instead of fish.

In a sanctuary, captive orcas can explore, swim, dive or be alone in the safety of a space in the natural world.

WATCHING A NEW SPECIES EVOLVE

Although we don't know when and how all of these cultural differences emerged, we now have a front-row seat to what some scientists believe is the evolution of two separate species of orcas, as the SRKW and transients continue to diverge in behavior and in biological adaptations. Remarkably, these adaptations are all based on behavior choices and learned cultural traditions adopted by each orca community. Most people think that

For orcas, each breath of oxygen is a conscious act. This orca pops up to take a breath and enjoy the sunset.

animal biology shapes behavior. But this is an example of behavior actually shaping biology!

Different orca groups across the globe, called ecotypes, express different behaviors throughout a wide range of open ocean and coastal habitats. Each population represents a different culture with a distinct dialect, prey specialization and hunting strategy (many hunt cooperatively, like wolves).

Each community of these big-brained, intelligent, emotional and socially complex mammals is unique. Each pod has its own social structure, and each individual orca is irreplaceable.

We know about orca lives from decades of observing them in their natural habitat, coming to know each as an individual, and understanding how that individual relates to other members of his or her social group.

And we know, from studying the orcas who have died, that their brains are over twice as large as expected for their body size, are very complex and the most convoluted on the planet, having more wrinkles on the surface than even humans. (Those wrinkles or convolutions indicate how much neocortex, the higher-order-thinking part of the brain, has become elaborated over evolutionary time.) The fact that orcas have a more wrinkled cortex than humans has thought-provoking implications for how intelligent—or "brainy"—they are. —*Lori Marino*

Lori Marino, PhD, is a neuroscientist and president of the Whale Sanctuary Project. She has studied dolphin and whale brains and intelligence for over 30 years.

Born to Be Free

The big brains, wanderlust, close bonding and cultural customs of orcas point to one conclusion: They cannot flourish in concrete tanks in marine parks.

Whether performing or not and whether born in captivity or captured from the wild, orcas in display tanks are deprived of everything they need—space, an interesting environment, challenges, continuity in their family and social life—not to mention a free choice about how to spend their days.

The chronic stress of living in tanks weakens the animals' immune systems and leaves captive orcas vulnerable to well-known opportunistic infections such as pneumonia, encephalitis, gastric disorders and candidiasis. These illnesses shorten their life spans considerably, with few living past the age of 25. (Most live at least twice to three times as long in the wild.) They also exhibit behavioral abnormalities such as endlessly swimming in stereotyped circles, remaining suspended on the surface of the water, self-harming and exhibiting hyperaggression.

The most well-known case of abnormal aggression is that of Tilikum, the subject of the influential 2013 documentary *Blackfish*. Tilikum was an orca captured from his homeland in Iceland at 2 years old and transferred in and out of various display facilities over the years when, in 2010 at SeaWorld Orlando, he killed his trainer, Dawn Brancheau. Even before the highly publicized incident, Tilikum had killed two other people in other facilities. In 2017, at the age of 35, he succumbed to bacterial pneumonia.

Although there have been hundreds of cases of captive orcas attacking their trainers, there has never been a single incident of a free-ranging orca seriously harming, let alone killing, a human or another orca. This tragedy underscores the abnormal psychological state of orcas that are forced to live in captivity in concrete tanks. Despite the fact that captive orcas endure physical and psychological harm, there are still more than 20 orcas in entertainment parks in North America and about 30 to 40 more captured orcas held in tanks in other countries around the world.

Moreover, two orcas in North America continue to live alone, without another member of their species. Lolita at Miami Seaquarium is the last surviving member of the Penn Cove culls. She has lived without another orca since 1980 in a tank only four times her body length with two captive Pacific white-sided dolphins. Her previous orca companion, Hugo, died of a brain aneurysm in 1980 after repeatedly slamming his head against the tank walls. Kiska at MarineLand Canada has lived alone since 2011, having lost all five of her calves while they were still juveniles.

The high mortality rate is a grim indicator that, despite being provided with regular food and veterinary care, captive orcas die more often and earlier than those who live in the wild. Surviving means a quality of life one would not wish on any living creature, animal or human.

While it might be tempting to consider releasing these captive orcas, experts warn that, too, would be disastrous. Most are either born captive or have been in the tanks so long that they now lack the necessary skills they would need to survive on their own. They do not know that live fish are food and have no family to help them adapt. They have missed the

Orcas in tanks are so stressed, most of them grind their teeth down to the nubs against surfaces.

critical period for learning skills to survive in the open ocean.

But there is an alternative that is part of a growing movement across the world—sanctuaries. There are successful sanctuaries for elephants, big cats, bears, great apes and many other species that are rescued from zoos, aquariums and circuses or that may have developed a life-threatening injury. An authentic sanctuary is a place where the well-being and independent choices of the residents is the priority—not the ticket sales that the animal may generate. And in a more humane environment, individuals can indeed flourish.

A handful of cetacean sanctuary projects are now underway around the globe. In 2016, I founded the Whale Sanctuary Project, a U.S.-based nonprofit organization whose mission is to create a permanent seaside sanctuary for captive orcas and beluga whales. The first of its kind in North America, the sanctuary will be in Port Hilford, Nova Scotia, a beautiful cove that will provide a permanent natural home for about eight whales in an expansive 100-acre netted area with lots of depth for diving and other aquatic animals to interact with. The residents will be fed and provided top-quality veterinary care but will enjoy the freedom of making their own choices about what to do on a daily basis for the first time in their lives. To find out more, visit whalesanctuaryproject.org.

—Lori Marino

A prairie dog views the terrain at Canada's Banff National Park.

When prairie dogs meet, they kiss to decide if they've encountered a friend or a foe.

How Social Networks Make *Animals Tick*

There are some universal truths about how animals interact. Even before Facebook, they formed lifelong bonds and navigated complex communities.

Jennifer Verdolin, an animal behavior researcher, has spent hundreds of hours watching prairie dogs lock lips. It may sound like a crazy pursuit, but this study of kissing rodents may end up teaching us a lot about how animals interact with their environment. This is how it works: When two prairie dogs approach each other, they may lock teeth and "kiss."

While it may sound sweet, the kiss is really a way of figuring out who is a friend and who is an enemy. "It can be a sign of who's in your group and who's not," says Verdolin. "If they belong to the same social group, they kiss and part ways." If they don't, the animals break apart and fight or chase each other away.

Prairie dogs were a natural group to study, because they have complex social networks—and lives that are almost as complicated as humans'. The rodents live in underground colonies composed of up to thousands of individuals. Each colony can be further broken down into groups, usually made up of an adult male, several adult females and their babies. And like humans, conflicts often arise within those larger communities.

Verdolin collected data on three Gunnison's prairie dog (*Cynomys gunnisoni*) colonies in Arizona, ranging in size from 60 to 200 individuals. Each colony is made up of numerous, smaller social groups consisting of three to 15 prairie dogs. The researcher painstakingly logged the animals' actions: who met with whom and what happened when they met. Individual prairie dogs were marked with microchips and black semipermanent hair dye so they could be identified. Verdolin collected a wide variety of social data, then specifically focused on kissing interactions.

Verdolin had the sense that certain prairie dogs were more likely to be super-social, but she wasn't entirely sure. One day, she was presenting her data when another researcher, Amanda Traud, had a realization. Mapping out these kisses wasn't just a way to understand animal behavior: The data itself created a complex social network.

INSIDE MOVES

Traud, who now runs a data-science firm in Washington, D.C., developed statistical tools and computer models using social network theory to analyze one year's worth of data on the three colonies and figure out which animals

Four mountain lion cubs groom with their mother at Masai Mara National Reserve in Kenya.

were interacting. She found that Verdolin's hunch was correct: There were individuals among the prairie dogs that were indeed hubs of activity inside the groups. These hyperconnected socialite prairie dogs had a disproportionate effect on the movement of information, food and disease through the network.

The study of social networks can help identify features of species that are invisible based on studies of individuals or behaviors alone across groups in all social species, from microbes to humans. "There was no way that I was able to tell this substructure or see the cliques" that were found through analysis of the social networks, Verdolin says.

In the case of prairie dogs, the networks can be vital to conservation efforts. That's because prairie dogs suffer from a disease called sylvatic plague—a kind of bubonic plague—which kills up to 90% of infected animals. If prairie dogs have to be moved because of an outbreak, it's crucial to identify which individuals belong to what groups. For example, there's a possibility of slowing or even stopping the spread of the plague in a colony by relocating these bridge individuals, the researchers say. At the same time, conservationists could use this data to be sure to relocate all the prairie dogs in a relevant social group, rather than splitting up groups. That could improve the group's chances of thriving in a new environment.

When prairie dog mother and offspring kiss, it maps their special relationship within the context of the wider social group.

NETWORKED SPECIES

And it's not just prairie dogs. Social network analysis can help conservationists reintroduce species to an ecosystem. Researchers in Zambia recently tapped social network data to help save lions. They assessed a captive pride's social structure and the relationships between individuals. Using data on how the lions played, groomed and greeted each other, the researchers tried to create groups for reintroducing the animals into the wild.

The same approach has offered some universal truths about animal

Before a lion is reintroduced to the wild, researchers study the individual's personal traits to find a pride where it would fit in best.

interactions. As groups get larger, there is a natural break: The groups become more fragmented and form cliques. And the more food that becomes available, the less likely the animals are to be in close association with each other, Verdolin says. In dolphins, when food was less available, disparate groups were more likely to come together and work as a single unit. "When I read about information transfer and eavesdropping in online networks, all I can think about is how that happens in animal networks too: They steal food or spy on other animals when they are hiding their stash."

Network analysis also helps detectives catch hackers and can help human resource departments identify job candidates. Traud is currently using a network-analysis approach to figure out why some hospitals have a high rate of neonatal intensive-care-unit admittance and why people use the emergency department for medical issues that don't require emergency care.

There are still questions to be answered in animal social networks. For one, what happens when key individuals are removed from a group: Does a new individual take on that role? "Humans and animals are so similar; the science applied to both can give us similar information and a better understanding of our relationship to the natural world," Verdolin says. "It's fascinating for people to know that the same tools we use every day—Facebook and the like—are poised to answer big questions in health, conservation and science."

Connected Lives

Humans are far from the only species with complex social lives. Here are other creatures that form networks and what we can learn from them.

Bottlenose Dolphins
A study of bottlenose dolphins in Indian River Lagoon in Florida showed that dolphins don't interact with all other dolphins to the same degree—they choose small groups with which to live and work. In particular, the researchers found that dolphin communities occupying the narrowest stretches of the river had the most compact social networks—similar to humans who live in small towns and thus have fewer people with whom to interact. Other research shows that each dolphin in a community has his or her own unique signature whistle.

Ants
Ants utilize a system of chemical signals to communicate, and they rub antennae together to transfer secretions to each other. A recent study found that not all ants are as social as others: Some ants communicate with only a few fellow ants, while others are social butterflies and communicate with a much larger circle. Researchers watched these interactions and found that on average, each ant had interactions with about 40 other ants. However, about 10% of the ants were especially connected, making more than 100 contacts with other ants.

African Elephants
African elephants are some of the most social animal species on the planet. They live in societies where social groups divide and re-form over the course of hours, days or weeks as the groups respond adaptively to changes in the physical and social environment. At the core of these relationships is a mother and her offspring, though groups also include independent males and even strangers at different times. Researchers are applying their studies of social networks to help them create more groups of elephants in captivity and the wild.

A mother elephant teaches her baby at the Masai Mara National Reserve in Kenya.

Mother and offspring form the hub of social life in African elephant communities.

When *Animals* Grieve

We've long maligned animals as unfeeling and unaware, but new research suggests that creatures from elephants to crows are alarmed by death and mourn the dead.

Cats do miss those close to them. About 46% of felines show lowered appetite after the loss of a companion and may either sleep more or suffer from insomnia.

"Animals have these advantages over man: They never hear the clock strike; they die without any idea of death; they have no theologians to instruct them; their last moments are not disturbed by unwelcome and unpleasant ceremonies; their funerals cost them nothing; and no one starts lawsuits over their wills," the French philosopher Voltaire wrote to Count Schomberg in 1769.

These days, of course, we know (as Voltaire did not) that we share an evolutionary history with all members of the animal kingdom. There's also increasing evidence that because of our shared biological past, we have similar behaviors, mental processes and emotions. Yet many still believe, as did Voltaire, that our human knowledge and understanding of death separates us from other animals. We are the only ones, these people say, who know we will die—that one day, slowly or suddenly, our life and all our hopes, loves and dreams will end. Surely, it is this awareness that sets us apart from the animal kingdom. Look at some of our greatest art, music and literature—all inspired by the certainty that death awaits us and every living being.

But some researchers think that we are giving ourselves too much credit. Other animals, they say, also grasp life's limits—and, like us, they mourn the loss of those they've loved. How else to explain the numerous sightings of whales and dolphins carrying their dead calves in their mouths or on their snouts or backs—sometimes for days? Often, other members of their pods trail the grieving parent, like mourners in a funeral procession (many of these can be seen on YouTube). Fishermen and scientists who've witnessed firsthand such heart-wrenching scenes say it's clear the adult bearing the dead youngster is in pain and distressed—as any parent would be.

When 9-year-old chimp Thomas died, 43 chimp companions touched and smelled him, lingering for 20 minutes.

FEELING THE PAIN

An ape's capacity to recognize death and suffer emotional loss seems apparent in the case of Thomas, a 9-year-old chimpanzee who died at a chimpanzee orphanage in Zambia in 2010. Researchers filmed the reaction of 43 of the other chimpanzees; 38 gathered around his body and stayed with his remains for nearly 20 minutes. Some of the apes touched or smelled Thomas' body or simply studied him closely.

One visitor, a chimpanzee named Masya, arrived carrying her own dead infant (a respiratory illness had swept through the chimps' community at the time). She'd previously been seen placing her dead baby girl in the sun on a patch of soft grass; then she retreated to the shade and watched the carcass. Every few minutes, she rushed to the body, as if she'd detected a stirring. She leaned in and intently studied her baby's face, peered into her open mouth and wide eyes and brushed away the flies. Finally, she placed her knuckles against her infant's neck—hoping, it seemed, for any sign of life. The chimpanzees at Thomas' side also appeared to be trying to come to grips with what had happened, though it seems likely they understood from his scent alone that he was dead.

Barely a decade ago, scientists discovered that many insects, from cockroaches to ants to webworms, recognize death's odor—a chemical cocktail of oleic and linoleic acids that individuals emit when dead. Insects that encounter it scramble to get far away. Ants even have cemetery areas where they deposit their dead. Once,

Did You Know?
Ants (like other insects) recognize the scent of death, and when they detect it in another ant, scramble to get as far away as possible. The carcass of a dead ant will be left for two days, then removed and deposited in an ant cemetery.

Two cows nuzzle in the field where they live. Cows are seen as plodding herd animals, but they actually have complex emotional and social lives.

famed ant researcher E.O. Wilson put a single drop of oleic acid on a worker ant. She was seized at once by fellow ants and carried "kicking and screaming" to the cemetery.

Of course, we humans also recognize death's smell. Yet at times, we simply refuse to accept the cold fact of that odor. We don't want to bear our loved one to the cemetery.

Similarly, one chimpanzee who'd been a close friend of Thomas seemed unable to handle the scene. He left the group but returned several times and stepped between others to get as close as possible to the body. Finally, he erupted in frantic screams while walking rapidly over the corpse. Was this a chimpanzee wail of grief? The researchers were reluctant to use such human-laden terms, yet we have no other vocabulary for this behavior. As Charles Darwin once wrote, "Who can say what cows feel when they surround and stare intently on a dying or dead companion?" It may be that we are the only animal with foreknowledge of death, but when it comes to grieving, we are not so unique.

EMPATHETIC MINDS

The study of animal grief is a relatively young field, largely because any animal behaviors that looked to be "human" were ignored for much of the 20th century. Scientists were worried that they might be mocked for anthropomorphizing—regarding animals as if they were people in fur or feather coats. Animals were also thought to be largely reactive beings that lacked thoughts and emotions and responded to stimuli as unthinking, unfeeling robots.

Two-thirds of dogs show signs of grieving after the death of another dog in the house.

It is impossible to know what dogs feel, but they have been known to wail at grave sites. Dogs experiencing grief often appear lethargic and lose their appetite.

That view has changed as wildlife biologists and others report on the emotional nature of other animals, recognizing that, like us, they also experience love and joy, not simply anger and fear. Grief is part of that spectrum, and researchers no longer worry that they will be viewed as sentimentalists for reporting such events. Indeed, in the past few decades, biologists have amassed so many firsthand accounts of animals caring for and mourning their dead that the idea of animal grief is no longer suspect. Such sorrow is not only limited to cetaceans and apes but also is seen in rabbits, goats and turtles. Biologists have even measured the hormonal reaction in wild baboon females who lost an infant or other close relative. Initially, their stress hormones spike (which can inflict microdamage on the brain), and they spend most of their time alone. Eventually they seek more companionship, and their hormones subside, leading the researchers to speculate that grief may have an adaptive value: It allows the animal to recover from this minor brain trauma and carry on. Indeed, all the baboons recovered from their sorrow, made new friends and gave birth again.

"Grief occurs widely in other social mammals and in birds—for example, after loss of a parent, an offspring or a mate," British psychologist John Archer wrote in his 1999 study, *The Nature of Grief*. It occurs in all animals—humans, included—because we all love. Love may be an ill-defined emotion and a strange force, yet we can no longer deny that other animals experience it too.

Love, elephant researchers say, explains the empathetic and caring behaviors of these great mammals when confronted with an ailing or dead companion. They may try using their tusks to lift up sick individuals—even dead ones—sometimes snapping an ivory in the process. Elephant calves have been seen standing forlornly, like tiny sentinels, at their dead mother's side. And like cetaceans, mother elephants may carry the corpses of dead infants with them. Most hauntingly, elephants investigate the remains of dead elephants they encounter, following an almost ritualized behavior.

"They stop and become quiet and yet tense in a different way from anything I have seen,"writes Cynthia Moss in *Elephant Memories*. "First, they reach their trunks toward the body to smell it and then they approach slowly and cautiously and begin to touch the bones, sometimes lifting them and turning them with their feet and trunks." The huge animals are particularly interested in the skull and tusks and gently run the tips of their trunks over these. "I would guess they are trying to recognize the individual," Moss writes.

When elephants meet up with companions they've not seen in some time, they trumpet their joy, spin in circles and pee buckets of urine. They clasp trunks, rub against one another and put their trunks in one another's mouths. "They know and love each other," Moss says. And so when death comes, they do what loving creatures do: They mourn. Grief is not ours alone; it is a pain we all experience as members of the evolved animal kingdom.

Last Rites

Giraffes have been observed standing vigil over their deceased. For instance, in 2010 in Kenya's Soysambu Conservancy, researchers reported that a mother spent four days lingering by her dead month-old calf. That same year, in Namibia, a herd of giraffes reportedly spent days investigating the corpse of a dead female. Male giraffes even spread her legs to learn more.

Crows often place twigs and leaves beside a companion crow's dead body and gather in a group, cawing loudly around crow remains. They are so observant that if a particular human has been associated with the crow's death, that person will be shunned.

Cats Although perceived as emotionally aloof, some cats observe the passing of other cats by repeatedly visiting the place where a deceased cat friend or companion once lived.

Elephants cover the remains of a companion with sticks and dirt. They can stand by the body of a herd member that has died for up to a week and appear to pay homage to elephant bones—but not those of other animals—they meet along the path.

Mother with son: Only about 4,000 tigers are left, but there's hope because a tigress can have up to 15 cubs in her lifetime.

The Real Tiger Moms

This female cat is committed, caring and ruthlessly protective when raising her young.

In India's renowned reserves, it's relatively easy to see one of the most iconic animals on earth—a Bengal tiger. These majestic, confidant cats are the kings and queens and the reigning predators in the Asian landscapes they inhabit. They are massive and muscular, possessing fearsome claws and teeth and a roar that resounds for miles. They radiate power and inspire awe. I've seen them lounging, sleeping, walking along and ignoring a lineup of safari jeeps packed with tourists, acting as if no one was watching them.

But tigers are also mysterious and secretive. They pad unnoticed on silent paws in the murky dusk and dawn and in the dead of night, the slashes of their black stripes concealing them in tall grass and forests. They seemingly appear out of nowhere, like apparitions. It's no wonder that tribal cultures have long deified this cat, bestowing powers beyond those of any worldly animal.

Tigers are particularly guarded in their social lives, living a solitary existence in carefully demarcated territory. They come together only briefly to mate, or for a female, for the years she spends raising her cubs.

The large cats communicate by spraying, rubbing or scratching on trees or in the dirt, claiming territory as a way to prevent encounters with rivals that could prove deadly. Other messages invite a romantic liaison. Tigers send these dispatches via scent glands located on their heads, beneath their tails and on their feet.

ON THE TRAIL OF CUBS

It's during the first few months of her cubs' lives that a female tiger is most guarded, a lesson that renowned big cat photographer Steve Winter learned while working on a tiger story for *National Geographic*. He'd heard from the director of central India's Bandhavgarh Tiger Reserve that a 4-year-old had just birthed a litter deep in the forest, far from the nearest road. It was too dangerous to walk there. You could only get in by elephant.

Winter hired E.A. Kuttappan to guide him. He was India's most famous mahout—an elephant handler who headed the anti-poaching patrol that tracked the tigers daily and searched for intruders. It was May, with temperatures soaring near 120 degrees Fahrenheit. To protect the elephant from the steamy heat, they had to leave camp before first light and return before midday. The elephant carried them through thick foliage until it finally opened into a narrow, rocky gorge.

And there was the tigress, regally lying on a ledge. She was about 8 feet long, not counting her tail.

"She'd returned to the same cave where she was born to have her first litter," Winter said. It was an ideal location. The cubs were denned above, out of sight amid a series of caves, giving her safe places to hide them from predators: hyenas, wolves, snakes or other tigers. With water nearby, she rarely had to leave—and could be the vigilant nurturer that she needed to be. Up to 50% of cubs don't even survive their first few months.

The tigress had known Kuttappan and his elephant since she was a cub, so she was completely relaxed around them. Her ears were up, her tail held high as she moved toward them. If she were afraid—or about to attack—her body language would have been glaring. She would have flattened her ears, bared her teeth, her tail lashing or held low—and she would have been growling or snarling. But she lay down and dozed. Wandered to the creek to drink. Sauntered over to a patch of sunlight and napped. There were no cubs to be seen.

If she had been concerned, she would have abandoned the site and hidden her babies elsewhere. But she stayed. Days passed and Winter occasionally glimpsed movement or saw the tigress vigorously licking what looked like a small tuft of fur. They sometimes heard the cubs crying and the tigress "speaking" to them in the soft groans she used to call or comfort them. There were only two mornings during the month that Winter photographed there that the female was elsewhere; on one of those days, he spotted her feeding on a deer kill nearby.

No two tigers have the same stripe pattern—like human fingerprints, they are unique to each individual.

READY TO LAUNCH

It ultimately took 24 days for one of the skittish cubs to finally emerge. It appeared only briefly to nurse, came up and rubbed faces with its mother and then disappeared. Winter got his photograph—and over the next days didn't see another cub. The skittish behavior of the cubs may have saved their lives. It would be a long time before they could protect—or feed—themselves.

Tigers may be apex predators and the largest of all the big cats, but females are nurturers that spend up to two years raising their young.

Winter followed another tigress in Bandhavgarh that had three nearly grown cubs. They were 17 months old and almost as big as she was. Though she'd been teaching them to stalk and ambush since they were about 6 months old, she still did most of the hunting for them all. Tiger expert Valmik Thapar describes the female tiger as a "committed, caring and ruthlessly protective" mother. And indeed, they are. *—Sharon Guynup*

Sharon Guynup is a National Geographic Explorer and a fellow at the Woodrow Wilson Center in Washington, D.C. She is co-author of Tigers Forever: Saving the World's Most Endangered Big Cat.

Tigers are not track stars:
Rather than chasing their
prey, they stalk and lie in
wait until they can ambush.
Some cultures
thought that
tigers carried
supernatural
messages.

A bonded gibbon couple spends hours every day touching, hugging and grooming each other.

Gibbon couples often "sing" together to help stake out their territory.

Soul *Mates*

Many of our primate ancestors formed committed couples and have continued to do so through millennia. The question is, why?

We may have to rethink the phrase "monkeying around," with its suggestion of promiscuous behavior. It turns out that many species of nonhuman primates have long been swingers only from branch to branch, not from partner to partner. In fact, they pair up for life, often monogamously, despite having none of the cultural and religious strictures that guide so much of human life.

Most primates, such as bonobos and chimpanzees, live in multimale, highly organized hierarchical societies. About one-third of primate species, like the orangutan, live solitary lives, with males mating on the run, periodically intermingling with groups of females. But about one-fifth of nonhuman primates, from owl monkeys and coppery titis to marmosets and tamarins, live in pairs.

Different human societies have also engaged in all three lifestyles, but pair living predominates in human culture as the socially acceptable standard, popularly believed to set us apart from other species. However, research shows that many of our primate ancestors beat us to it. Primatologist Peter Kappeler, PhD, of the German Primate Center, and anthropologist Luca Pozzi, PhD, of the University of Texas at San Antonio, have attempted to explain the mystery of pair living, not just now but in our evolutionary past.

DUELING THEORIES

Scientists have long wondered how and why pair living came to pass, since it would seem to be counterproductive for the male, whose natural biological imperative is to spread his seed as far and as wide as possible. Settling with just one female who has to go through long periods of pregnancy and maternal care of the young would run counter to that. "Living as a pair represents a puzzle in the evolution of mammalian social systems because males could achieve higher reproduction if they did not stick with a solitary female," says Pozzi.

To try and solve the puzzle, Kappeler and Pozzi analyzed the genetic data and typical behavior of 362 different species of primates and constructed a model of how their social relationships may have evolved over time. They specifically explored two central theories behind pair living. The first theory, known as "female spacing," holds that when food and other resources in one area run low, females tend to spread out. "When they do so, they are less likely to have to compete with large groups of females, less likely to run into large groups of males, and more likely to pair up with one," Pozzi says. The second theory, called "the paternal care hypothesis," posits that in some nonhuman primates, just spreading their seed isn't enough for males; they gain an evolutionary benefit from pitching in and caring

Different lemur species display a variety of sexual systems, from monogamy to polygamy.

for offspring alongside mom. "Under certain ecological conditions where there are not enough resources, pair living allows the couple to pool their resources, finding food together, defending their territory together and making sure their offspring survive," Pozzi explains.

The female-spacing theory and the paternal-care theory have always been dueling hypotheses. But Kappeler and Pozzi found otherwise. "For a long time, these theories have been commonly seen as mutually exclusive, with support for one or the other," says Pozzi. "But based on our research, we suggest they go together. We found that female spacing may help in creating the pair unit and paternal care establishes a much longer curve for it."

In many species like gibbons and owl monkeys, the relationship is largely sexually monogamous, says Pozzi. Gibbons, in fact, actually sing operatic if cacophonous duets together, loudly trumpeting their commitment to one another. (If you'd like to hear the unique mating song from the gibbon world, go to youtu.be/1PZC1Sz9XFM.) Other species, such as the lemur and tarsier, are socially but not sexually monogamous, periodically finding outside reproductive mates, but still sticking together or coming back and raising their children together as partners for life.

A mother wolf cuddles her son: Wolves have among the tightest family units in the animal kingdom.

THE KINSHIP HYPOTHESIS—OR WHY SISTERHOOD MATTERS

As the environment changes over many generations, perhaps as resources become more plentiful, some species of paired couples learn to share their spaces communally with others and thrive together. But no species goes straight from single living to group living—pair living is the bridge. "In our model, we see that it is almost impossible to go straight from solitary living to group living," says Pozzi. There are variations on the theme: Some species, like the orangutan, stay solitary; others go on to and stay in pairs; yet others go to group living whenever resources are flush, but then return to living in pairs when resources become scarce again. However, none have gone straight from solitary living to group living.

"Pair living appears to be the necessary stepping stone to more complex societies," Pozzi says. Just how the transition from pair living to group living evolves among primate species is another story, but Pozzi notes that their findings support yet a third theory: the "kinship hypothesis."

In some circumstances, he explains, after a male and female have enough daughters, small groups begin to form, thanks to kinship among the females. Male offspring are likelier than daughters to move away and form new

Like many devoted primate couples, beaver pairs stay together "as long as they both shall live."

pairs or groups, but the females have closer relationships and tend to stay.

Living close together, these related females will welcome one or more males into the group. But this happens only if the environment's resources are plentiful and the competition between females in the group is low, Pozzi explains.

Among primates, females tend to drive the structure of the group. "It's literally a sisterhood, with cooperation among kin," Pozzi says. In some species, such as gorillas, the females frequently end up forming a kind of harem together, sharing one male and keeping the other males out. In other cases, as with the chimpanzees, more males are invited in, and multiple males and multiple females share one another, creating larger, more complex societies, often complete with hierarchies, including alpha and beta males.

HOW HUMANS LEARNED TO LIVE IN PAIRS

While our primate ancestors appear to have preceded us in the practice of pair living, humans go about it in a dramatically different way, Pozzi says. In fact, we followed a reverse route, going from groups to pair societies rather than the other way around.

"As with earlier primates, a lot of things, including biology, should have led us not to be monogamous," he says. For humans, pair living and monogamy arose from cultural rather than biological factors, Pozzi notes.

In early, pre-agricultural hunter-gatherer societies, we were not monogamous, he points out. "Just

Marmoset couples groom each other for hygiene, social contact and conflict resolution.

Marmosets live in families of three to 15 individuals and are socially monogamous.

like our closest ancestors, such as the bonobos, chimps and pygmy chimps in Central Africa, we were highly promiscuous in multimale, multifemale relationship groups, and like those earlier primates, we had sex not as much for reproduction as for social bonds."

Then about 10,000 years ago, agricultural societies came along, and we started having to deal with property. "One of the major problems for human males was to be sure about paternity, so that you could transfer your property to the next generation, and this was a catalyzer for pair living and monogamy," says Pozzi. Religion and cultural norms built upon this foundation after that.

As in modern human society, males in bonded primate couples take part in all aspects of rearing the young.

But as we all know, human couples stray, divorce, marry again. Some have open marriages and some, like certain Mormon sects, are polygamous. And some hunter-gatherer societies that still exist today, like African pygmies, have more relaxed societies with routine promiscuity. For all our religious and cultural trappings, human pairing and sexuality are no more fixed in stone than they were for many of our primate ancestors.

The Biology of Love

Bonding takes place in the brain. A few mammal species have special circuitry compelling them to stick with one partner for life.

Primates aren't the only mammals who pair up for life. In fact, 3% to 5% of mammalian species, including wolves and rodents, such as beavers and voles, tie the so-called knot. The question is, what bonds these species together as duos when the vast proportion of mammals play the field?

Behavioral neuroscientists at the University of Colorado, Boulder, think they may have found an answer, and it brings to mind the old adage "absence makes the heart grow fonder." It turns out that these more monogamous species may be programmed for longing—if they've been separated, a certain center in their brain fires up like a lighthouse when they get back together.

The research team, led by neuroscientist Zoe Donaldson, PhD, observed the behavior and monitored the brain activity of a species of monogamous rodents called the prairie voles—which resemble a combination of a hamster and a lemming—in an effort to gauge what brain regions drive their instinct to form lasting unions. Using tiny cameras and a new form of brain imaging called in vivo calcium imaging (which can home in on individual neurons in living animals), they spied on the brains of dozens of voles when they were meeting for the first time, then three days after mating and, finally, 20 days after they had moved in together. They discovered that the nucleus accumbens in their brains—the same area that lights up in hand-holding human couples—consistently started glowing when they came together after a separation. The longer the couple had been together —the tighter their bond—the larger the cluster of glowing cells in their brains. And when the lovers touched, the signals flamed all the brighter. This blatant turn-on in their nucleus accumbens did not occur when they were approaching vole strangers.

The researchers hypothesized that a comparable "partner approach response" also happens in humans, and they hope to study this in future research projects. "There has to be some motivation to be with a specific person and maintain relationships over time, and ours is the first investigation to pinpoint the potential neural basis for the urge to reunite," said Donaldson. "We are uniquely hardwired to seek out close relationships as a source of comfort, and this often comes through physical acts of touch. We have a neuronal signal telling us that being with loved ones will make us feel better."

Donaldson suspects that feel-good brain chemicals such as oxytocin, dopamine and vasopressin, which are known to have a role in fostering trust and closeness, are involved. She also believes the study may shed light on why social distancing (as in the wake of the COVID-19 pandemic) can be so hard for us.

"It's the equivalent of not eating when we're hungry, except instead of skipping meals, we're slowly starving emotionally," Donaldson says. If the research pans out in humans, she adds, it might one day be tapped into to help treat autism, depression and other mental health disorders in which people have deficits in making social connections.

Honeybees carry pollen from male to female flowers, sustaining life.

A honeybee visits between 50 and 100 flowers while collecting nectar.

The Buzz *on Bees*

A beekeeper profiles the gentle community of pollinators fighting to survive.

From my kitchen window I can see a beehive. It's been snowing all day, and the snow is piling up on this hive like frosting on a cake. I wonder how the bees are. If they're warm enough. If they have enough honey to eat. If they're healthy. Before the cold weather hit they seemed content. But the cold struck early, and I'm worried.

Honeybees captivate me. These tiny beings are capable of creating honey and building cathedrals out of wax. Along with other pollinators, they are responsible for transferring grains of pollen from male to female plants, enabling production of at least a third of all food eaten in the world.

The honeybees' great skills of navigation have made them darlings of agriculture for centuries, and their method of communication still overwhelms me when I think about it: They communicate, in part, by dancing. Karl von Frisch, a Nobel Prize–winning scientist from Germany, showed that bees are so incredibly attuned to each other's body movements that one bee returning from a blossoming field can dance a map for other bees to those specific flowers, and she can do it in the dark of the hive! The bee shakes her body and dances in a line that corresponds with the angle of the sun. The other honeybees read her movements with their sensitive antennae. They can then go visit those same flowers. What has moved me most, though, is the gentleness and incredible care honeybees exhibit toward each other in the course of their lives.

FALLING IN LOVE WITH COMPLICATED LIVES

My love affair with insects began when I was a little girl. Spending most of my days on a prairie with my dog, I became fascinated by butterflies, grasshoppers, beetles, fireflies and bees of many shapes and sizes. The fields quivered with the business of their complicated lives.

When I was 8 years old, my father took me to visit his beekeeping friend during a honey harvest. We entered a shed filled with many frames of honeycomb. The smell of honey and wax filled the air with an unmistakable perfume. I was intoxicated. But what also struck me in that moment was the way that this beekeeper was so calm in the midst of all those bees flying around him, landing on his body without

In some regions, 90% of honeybees have disappeared from sight.

A new Brazilian study finds pesticides—even those considered to be "safe"—may reduce bee life spans by 50%.

stinging him. These fascinations led me years later to pursue beekeeping myself, not for profit, but for love.

Once I started beekeeping, I realized how truly extraordinary these insects are. Scientific studies of their behavior have taught us bits about how and why they live so harmoniously in community. Entomologist Tom Seeley, for example, has proven that honeybees make their decisions democratically. His work overturned the long-standing myth that a colony of bees was a monarchy and that the worker bees were only serving the wishes of the queen bee. In fact, they decide things by consensus. But I also learned that the honeybees—revered, studied and admired by cultures around the world for centuries—were dying.

In 2006, news flew around the globe about honeybee and pollinator decline, depressing me deeply but also compelling me to learn more about how to save them. I found there was no simple solution for the honeybee's plight. The problems included habitat destruction, ubiquitous pesticide use, new viruses and mites, and monocropping—the practice of planting giant fields of one single plant, like corn or soybeans. Because the bees used for monocropping are exposed to just a

Did You Know?

The ladybug was once a common sight across the yards of North America, but now these insects are relatively rare here. At the same time, ladybug species have expanded their numbers—and their range—around other parts of the world.

Honeybees care for each other: They groom their neighbors, sacrifice for companions and share their work.

single species of crop, when they finish collecting nectar, they've exhausted their food supply and can starve.

Unfortunately, some of these issues are not only a problem for honeybees. Insects of all kinds are being decimated. In the fall of 2017, a landmark German study showed that flying-insect populations had fallen by 75% in the past 25 years. Other studies have mirrored the news that insects are in trouble. Native bee populations in the U.S. have declined by 23%. Monarch butterfly populations have decreased by 80%. Fireflies are less abundant. And these are the more glamorous insects that have interested people enough to take notice.

Insects are the very foundation of our ecosystems. The disappearance of insects impacts bird populations, fish populations, amphibian populations—and, of course, the pollination of many kinds of plants. Interspecies cooperation is so complex. When we lose one insect, we may well be losing a bird or a fish that depends on that insect for food. The loss of a native bee might mean the loss of a particular plant. These facts fill me with horror, but I am also overwhelmed with sadness because honeybees—the insects I know most intimately—seem to experience emotions.

Honeybee colonies include a single queen and several thousand female workers and male drones.

EMOTIONAL CREATURES OF MANY MOODS

Most of the natural beekeepers I know would attest to the fact that hives display moods. In general, when the colony is healthy, the bees hum quietly and calmly. When the nectar flow is on during high summer, their energy is exuberant and zesty. At times when something is awry—a queen has died or the colony is not well—they will appear feisty or despondent. The existence of collective feeling inside the hive was intuited by Belgian playwright Maurice Maeterlinck in the early 1900s when, in his poetic book *The Life of the Bee*, he wrote: "They must be able, therefore, to give expression to thoughts and feelings, by means either of a phonetic vocabulary or more probably of some kind of tactile language or magnetic intuition, corresponding to senses and properties of matter wholly unknown to ourselves."

Over a century later, a study by ethologist Melissa Bateson and her team at Newcastle University in England proved scientifically that honeybees experience something akin to pessimism, and researchers at Queen Mary University of London recently did a study where bees exhibited optimism.

Personally, I have also witnessed bees suffering, limbs twitching uncontrollably after encountering toxins or moving with extreme lethargy when ill. We would never question a human experiencing pain if they were acting in this manner. I have also taken a bee that was out all night in the cold and appeared to be dead, warmed her in my hand and watched her gently touch my skin over and over once she was warm and mobile again. Was this an expression of gratitude?

CARING COMMUNITIES

Many beekeepers argue that science has not yet adequately described the many mysteries of honeybee relationships. But their capacity for what looks like emotion is not as

We cultivate honey after bees extract nectar from flowering plants.

Once a bee has a full load of nectar, she returns to the hive and passes it by mouth to other worker bees.

surprising to me as the way that they tend to care for each other.

Bees often groom each other in and around the hive. But recently I was struck by what looked like an unusually selfless moment of empathy. One sunny afternoon in autumn, when the honeybees who live in my yard were busy visiting meadows full of late-summer flowers, a sudden storm poured a brief and heavy rain over everything. As quickly as it came, it was gone, and the sun shone brightly again. Many honeybees had been caught in the downpour and their wings were too wet for them to fly. I began picking them up to carry them back to their home. When two of them met in my hand, they abandoned grooming themselves and began grooming each other. One had a torn wing but put the needs of her companion first. Taking care of each other: That's what bees do.

Other insects exhibit shared labor and collective care. In his book *Innumerable Insects*, bee taxonomist Michael Engel writes, "The insect world teems with protective mothers ranging from earwigs to leaf beetles and many of their behaviors are not that different from those of birds tending their eggs or feeding their chicks." The awareness of suffering, empathy and cooperation that honeybees and other insects exhibit is not often a convincing argument for people to care about them. The fact that insects feel alien to us has led us to think of them primarily as pests. And while there are insects that cause trouble for humans, our lack of empathy has allowed us to eviscerate their populations without questioning the larger impact. Maybe when we recognize how crucial they are, we will begin to change our destructive habits.

On a recent morning, I walk out to the hive near my house. Temperatures have already dropped below 20 degrees and I wonder if the bees have survived. I lay my head against the wall of the hive and listen. At first, I hear nothing, but I move my head and then I hear them. A quiet murmuring, as soothing as the chanting of monks. —*Heather Swan*

Heather Swan teaches environmental humanities at the University of Wisconsin-Madison. She is the author of Where Honeybees Thrive: Stories from the Field.

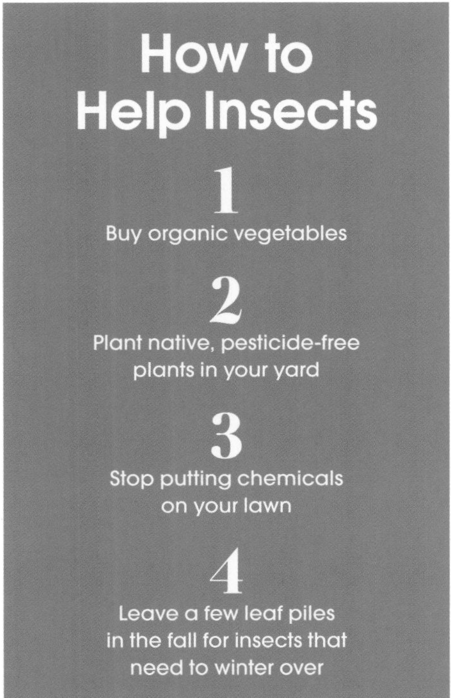

How to Help Insects

1
Buy organic vegetables

2
Plant native, pesticide-free plants in your yard

3
Stop putting chemicals on your lawn

4
Leave a few leaf piles in the fall for insects that need to winter over

What a Fly Feels

Once viewed as zombies in flight, insects now turn out to exhibit feelings such as optimism, depression and fear.

Fear of anthropomorphizing—comparing insects to humans—has kept many scientists from making claims about insect emotions. But by adapting the same nonverbal techniques used to measure emotions in human psychology, some researchers contend that insects do indeed seem to have emotional states. Studies measuring brain chemicals observe associated behaviors. In one, fruit flies are repeatedly startled by a hovering dark shape that looks like a predator and seem to exhibit fear and trepidation. Bumblebees given sugar rewards fly faster, an optimistic act, say researchers, indicating "positive emotional states" that correspond to higher levels of the neurotransmitter dopamine in the brain. Honeybees become pessimistic after being shaken and agitated. Ants display a wagging behavior that, in other animals, would align with expressions of pleasure—and these are especially pronounced when in contact with food or, for adults, with baby ants. There is much more to learn, but it could be that the insect world is alive with feeling and emotion, just like us.

A 12-year-old male chimp sits at the base of a tree in Bossou Forest, Mount Nimba, Guinea.

Walking With the Apes

Jane Goodall transformed the study of nonhuman primates and pioneered a new understanding of the animal mind.

"I wanted to come as close to talking to animals as I could."

In July 1960, a blond and lithe young British woman, Jane Goodall, stepped on the shore of Lake Tanganyika in the Gombe Stream National Park of Tanzania. She was there to launch a study of chimpanzees in the wild, a study no one had ever done before. Goodall had no training as a wildlife biologist; indeed, she hadn't even attended university. But she had a deep love of animals and a steady patience for observing them—and she was determined to succeed.

Goodall's unlikely beginning soon transformed our understanding of not only chimpanzees, but also animals in general, which led to the foundation of a new field of science, primatology (the study of wild, nonhuman primates). It also led to conservation efforts to protect the great apes and smaller primates in the wild. She arrived in that Tanzanian forest thanks to the efforts of Louis Leakey, at the time the world's leading paleontologist and expert on human origins. Leakey, together with his wife, Mary, had discovered the fossils of ancient early humans at Tanzania's Olduvai Gorge, proving that humans originated in Africa. Gifted with a rich imagination, Leakey thought the chimpanzees living along the shores of Lake Tanganyika were somewhat like those earliest humans, who had lived in a similar setting—although more than 2 million years ago.

Leakey met Goodall after she moved from the United Kingdom to Kenya in 1957 in the hope of finding some way to work with animals. She was staying

Jane Goodall plays with one of the chimps she studied at Gombe Stream National Park, Tanzania, in 1965.

Chimps laugh when they play. Mother chimps are attentive to their young.

A mother and infant chimpanzee cuddle in the grass.

at the farm of a family friend east of Nairobi, and they suggested she get in touch with Leakey at the country's natural history museum. He agreed to a meeting, where she explained to him her childhood dream: "Somehow I must find a way to watch free, wild animals living their own undisturbed lives—I wanted to learn things that no one else knew, uncover secrets through patient observation."

Leakey asked her many questions about animals, "and as I'd read all about them all my life, I could answer him," she noted. He was impressed that she knew the meaning of ichthyology (the study of fishes) and knew what ant bears were. "It wasn't just luck. He could see that I really was serious."

Leakey invited her first to spend three months working at Olduvai Gorge to see if she could handle living in the wilds of Africa. One night toward the end of her time there, she and Leakey walked to the edge of the gorge and he began telling her about "a group of chimpanzees that lived along the shores of a lake, very isolated and far away, and how exciting it would be to learn about their behavior." It was exactly what she longed to do, she told him. "Why do you think I'm talking about it?" he replied. If she would promise to devote herself to this research, he would find the means to send her to and support her research in the field.

IN THE LAND OF THE APES

Leakey had never been to the Gombe Stream National Park or seen a chimpanzee in the wild. He'd heard about it from a fellow Cambridge University anthropologist, who'd briefly visited the reserve in 1945. Leakey's lack of firsthand knowledge didn't deter him. Nor the fact that Goodall wasn't a trained scientist—or that she was a female. He regarded both as strengths. As a woman, she would likely be more patient than a man and less threatening to the male-dominated chimpanzee society. And her lack of higher education simply meant, from his perspective, that she possessed "a mind uncluttered and unbiased by theory."

It was two years before he found a sponsor, the Wilkie Brothers Foundation of Des Plaines, Illinois, whose founders ran a tool-manufacturing business and who were deeply interested in the history of tools. Leakey also received permission from the reserve's game wardens for Goodall and her mother, Vanne Morris Goodall, to live in the forest with Tanzanian assistants. Goodall's study got off to a promising start: On the day after her arrival, she spotted two chimpanzees close to their camp. They ran away. But she remained optimistic when she found that most mornings she could watch a large group of chimps feeding in a nearby tree. They didn't seem bothered by her presence. But 10 days later, they vanished, and for the next two months she didn't see another chimpanzee.

Chimpanzees communicate much like humans do, by kissing, embracing or patting each other on the back.

Determined to succeed, Goodall began following a daily routine. At the same time every morning, she would hike to an overlook and watch for chimpanzees. Often she spotted them in the treetops of fruiting trees, and she found that they would tolerate her presence if she did not move closer than 60 feet. Even from that distance, she could make out their differences: Some had the graying faces of elders; others the smooth, ivory-hued skin of youth; some had distinctive scars. Acknowledging their individuality, she gave them names: Flo, Goliath and Leakey. She made notes on their calls, sampled the fruits and nuts they ate, watched as they groomed each other and built sleeping nests among the fronds of tall palm trees.

Only four months after her arrival, she witnessed what no one had seen before: chimpanzees eating a wild piglet they had killed. They weren't vegetarians, after all! A few days later, she made another surprising discovery. David Greybeard, her favorite chimpanzee, sat next to a nest "carefully pushing a long grass stem down into a hole in the mound." She wasn't close enough to see what he was eating, but after he left, Goodall pushed one of his discarded stems into the hole and pulled out a cluster of termites. Later, she saw David Greybeard and Goliath making termite fishing tools from tree twigs by stripping off the leaves. This became one of her most important discoveries. Until that time, only humans were thought capable of designing and making tools. An elated Leakey sent her a telegram: "Now we must redefine tool. Redefine Man. Or accept chimpanzees as humans."

MEETING THE INDIVIDUAL CHIMP

As Goodall's work at Gombe became more widely known, Leakey urged her to study for a PhD. Despite her lack of an undergraduate degree, Cambridge University accepted her as a doctoral candidate in 1962. She published her first paper in a scientific journal the following year—discovering as she did the strong bias in the scientific community against her recognition of the chimpanzees as individual beings. The editor had changed her pronouns—he, she and who—to "it" and "which." She restored her original version saying years later that the final version "conferred on

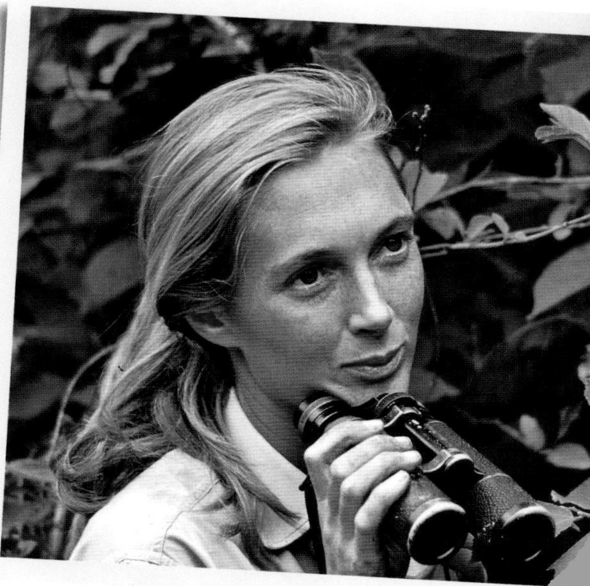

A young Jane Goodall studied chimpanzees from a distance through her binoculars.

Bonobos are closely related to chimps; both species share 98.7% of their DNA with humans.

the chimpanzees the dignity of their separate sexes." Still, other scientists objected to her naming the animals and her recognition of their individual personalities, arguing that she was engaging in "anthropomorphizing"—that is, treating the chimpanzees as if they were people in fur coats. They argued that it was more scientific to give these primates numbers, as is usually the case with animal studies in laboratories.

Goodall never retreated from her recognition of chimpanzees as individuals with distinct personalities, emotions and personal histories. In 1965, she received her PhD in ethology from Cambridge University. Thanks to an introduction from Leakey, Goodall also received grants to continue her study at Gombe Stream from the National Geographic Society. With the society's funds, she built a research station and small home close to the lake.

In 1964, she married Baron Hugo van Lawick, a Dutch nobleman, wildlife photographer and filmmaker, whom the National Geographic Society sent to film her at work. Three years later, she gave birth to their son, Hugo Eric Louis, known affectionately as "Grub." But the marriage did not last; she and the baron divorced in 1974. She later married Derek Bryceson, the director of Tanzania's national parks; he subsequently died from cancer in 1980. Now in her 80s, Goodall is the grandmother of two children.

A LIFE'S WORK

Goodall continued to study the chimpanzees at Gombe through the 1970s and '80s, although she

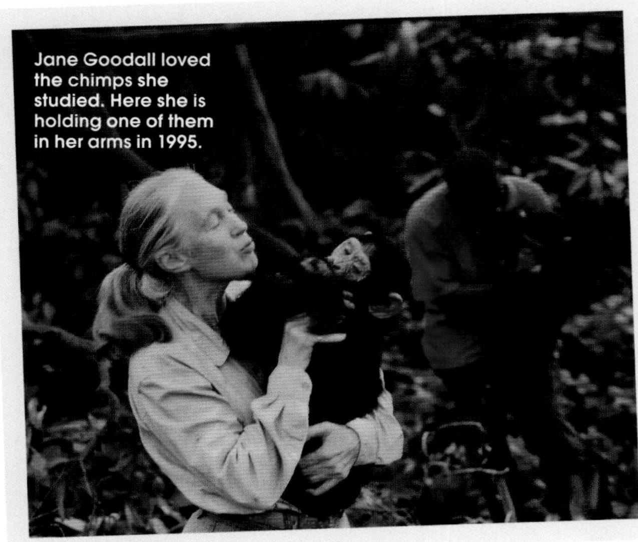
Jane Goodall loved the chimps she studied. Here she is holding one of them in her arms in 1995.

relied more on Tanzanian research assistants—and she discovered more startlingly human-like behaviors. She documented the chimpanzees' complex social system, including a "war" between rival groups; the use of social embraces to comfort a chimpanzee grieving the death of another; and the altruistic practice of adopting orphaned chimpanzees by others in the group. She has also argued that the chimpanzees show the beginnings of a primitive language system that includes more than 30 sounds with specific meanings.

In 1986, Goodall published *The Chimpanzees of Gombe: Patterns of Behavior*, a scholarly opus presenting in detail her rigorous research, which solidified her academic reputation. After a scientific conference organized around the release of this work, Goodall began to devote herself to conserving wild chimpanzees. The primatologist had already founded the Jane Goodall Institute in 1977 to promote health and conservation in the local communities located near chimpanzee habitats and started ChimpanZoo in 1984 to study and improve the lives of captive chimpanzees. She also started Roots & Shoots, a conservation program for young people and educators.

MESSENGER OF PEACE

In 2002, United Nations Secretary General Kofi Annan named Jane Goodall a U.N. Messenger of Peace, and in 2004, Prince Charles invested her as a Dame of the British Empire. These days, she still visits Gombe when she can. But she spends most of her time traveling around the globe, giving lectures and fundraising. She passionately urges her audiences to appreciate the power of what she recognized in the chimpanzees—the individual. "Every individual counts," she insists. "Every individual has a role to play. Every individual makes a difference."

All About Chimps

Much of what we know about these creatures comes from the legendary primatologist's work in Africa. Here, some fascinating chimp facts from the Jane Goodall Institute.

• Chimps are one of four types of great apes, which are: chimpanzees, bonobos, gorillas and orangutans.

• Chimpanzees use more tools for more purposes than any other creatures, except humans.

• In captivity, chimpanzees can be taught American Sign Language.

• One of the chimpanzee calls is known as the "pant-hoot." Each chimp has his or her own distinctive pant-hoot, so they can be identified with precision.

• Chimps groom each other. Grooming helps relations within the community and calms nervous or tense chimpanzees.

• When chimpanzees are angry or frightened, their hair stands on end.

• Like humans, chimps have opposable thumbs and opposable big toes, which allow them to grip things with their feet.

• Chimpanzees are not meant to be pets in our human homes; a full-grown chimpanzee has five or six times the strength of a human being.

• Chimps are endangered. There are probably between 172,000 and 300,000 chimpanzees remaining in the wild.

Wolf partners will collaborate to hunt and shelter and prefer to live their lives far from the human world.

The Alpha Dog Myth

Dogs are loving family members and friends, not pack animals to be subdued and beaten down.

"For the strength of the Pack is the Wolf, and the strength of the Wolf is the Pack," Rudyard Kipling famously wrote in *The Jungle Book*.

The idea of wolves living in packs of unrelated individuals with a powerful alpha leader at the helm has long been part of fiction and folklore. The nature of the wolf pack has been the subject of old and new research, too. And recently, the old myths have been debunked. It turns out that wolf packs are not actually a gang of strangers, but rather, biological families. The alpha wolf is the father, and the leading female, his mate. The other wolves, the so-called followers of the pack, are their offspring, in various stages of development. But who wants to let a little reality ruin a foreboding narrative? Along the way, not just wolves but also dogs were mixed into the tales of the alpha wolf. Eventually, facts and fiction blurred into questionable assumptions on canine behavior.

One result was the alpha dog myth. It goes like this: Dogs, the descendants of ancient pack animals with an alpha leader, the wolf, are also wired to be—or to respond to—an alpha leader. Although that belief is often repeated as fact, it's simply untrue, according to Leticia Dantas, DVM, PhD, a veterinary behaviorist at the University of Georgia School of Veterinary Medicine.

Go ahead, enjoy your large dog family. Everyone can chill because you won't have a leader of the pack.

The "alpha" concept in wolves took off after Swiss ethologist Rudolf Schenkel published behavioral research holding that wolves battled it out for the top social position, the alpha, "by incessant control and repression of all types of competition" within the same sex. But there was a problem: Schenkel wasn't studying wolves in nature, among their natural family members, but rather, animals thrown together in captivity, leading to flawed and mistaken assumptions. In reality, his work did little to reveal how relationships between wolves unfolded and were sustained in their natural habitat.

"Wolves in an enclosure are constantly under stress, are forced to interact, do not have opportunities to avoid each other (an important coping mechanism for all social species) and do not have social or affiliative bonds. They do not constitute a real pack or family group," explains Dantas.

Later, more rigorous scientific research found wolves in the wild do express dominance in wolf packs—but it's far from a wolf-eat-wolf existence.

Did You Know?
Dogs have a range of personalities that fall along a few discrete categories or types. Among the qualities, dogs may be confident or unsure, bold or shy, dependent or independent, laid-back or perpetually excited and ready to party on.

The relationship between you and your dog is driven not by social status but by reinforcement and love.

It takes nonviolent cooperation to maintain the group's cohesion and survival in order to hunt, travel and defend resources together.

After all, you can't reproduce or maintain your rank if you're a pack leader injured by constant fighting, Marc Bekoff, PhD, professor emeritus of ecology and evolutionary biology at the University of Colorado, explained in *Psychology Today*. Instead, much of the "fighting" seen in a pack of wolves—or any group of feral dogs—amounts to ritualized social signals, a way to "act out" using various postures to communicate threat, appeasement or submission without anyone actually going for the throat.

THE TRUTH ABOUT DOGS

So, how does this relate to dogs? The assertion that dogs live in strict hierarchies based on dominance with an alpha dog at the top isn't reality. It's simply not consistent with how dogs behave as a species, Dantas points out. "Studies of interactions between dogs show no evidence of fixed hierarchical relationships but rather fluid relationships based on learning," she says.

Nonetheless, the idea that pet dogs have strict pecking orders in groups has received some added fuel over the past few years from studies of feral dogs, research often erroneously applied to pet dogs. For example, over the course of a year, researchers from the University of Exeter studied a pack of 27 free-roaming dogs living in the suburbs of Rome. The canines were strays without any owners, but they relied on humans to leave them food. The research team concluded the pack had a hierarchy with older adult dogs expressing dominant behaviors over younger pack members while the younger dogs in the middle of the group were mostly aggressive with each other, especially about food.

Another study from the University of Parma observed the behavior of five packs of stray dogs over the course of months and came to a similar conclusion that hierarchies in the feral dog groups were common and related to age, with older dogs the most dominant. But a closer look shows these conclusions about dog hierarchies—and even what constitutes dominance in the wild—don't apply to companion dogs, even if several dogs are living in a household. Dantas points out that the types of studies cited above involve non-neutered animals who are not pets but living in packs where it's not unusual to find more aggressive interactions about resources—both food and mates.

"Resource guarding is not the same as dominance, but this confusion is all over the biology and ethology literature.

Another alpha dog myth: that a pooch running ahead considers himself the boss of the human left behind.

Dogs often sniff the genitals and anus of their friends to glean extra information. (Dogs do not realize this offends people.)

Dogs need a chance to stop and smell surroundings on their walks.

Free-ranging, intact animals should not be compared to our pets—a population that is highly socialized to humans to the point where it would be more appropriate for researchers to compare their bond with us to that of children," Dantas says. "Even their breeding is completely controlled by us."

Another factor that's contributed to confusion about whether a dog might be that mythical alpha is the difference in personality from one individual to the next. "Just like with humans or any other species, some individuals are more assertive, and others more insecure or shy. This is one of the factors that defines how relationships develop," Dantas notes.

Motivation and the emotional state in the moment play a role as well. For example, depending on how hungry a dog is, she's more likely to show teeth to her friend because she needs to eat right now. This is normal. Another factor is the history of the relationship between dogs—what happens as two individuals get to know each other will direct how their relationship is moving forward.

People who love animals will often adopt more than one pet over time. But Dantas urges dog lovers to remember the difference between a family you create and one born together. In nature, dogs are born into family groups where an emotional bond is formed from the start. "But no dog or cat decides to move in with a roommate. What we create as pet owners is an artificial situation where animals that did not grow up together all of a sudden have to share a space, resources, attention," she says.

Train your dog with delicious treat rewards, not physical violence or frightening threats.

"While some dogs are flexible and skilled enough to go along with it, others will struggle, and aggression can always happen when a canine is under stress," adds Dantas. If that happens, don't blame the problem on the alpha dog myth. Instead, opt for patience, kindness and, if needed, seek help for you and your pet.

The Alpha Human Myth: Cruelty Disguised as Training

Taking a positive approach to teaching your dog instead of establishing dominance will be more rewarding to both of you.

Consider these scenarios: One person is deliberately physically abusive to a dog, making it cower and shake. Another person claims to love his or her pooch but insists having a well-trained dog means the owner must be the "alpha pack leader" and dominate the canine with force, quickly punishing unwanted behavior.

Unfortunately, these examples are not as different as you may think. True, the dog abuser is obviously guilty of cruelty. But so is the person attempting to show the dog that a human must always be obeyed. Bottom line: If you've bought into the "I have to be the alpha dog" myth, it's time to give it up.

In the first place, you are not a dog—and your dog definitely knows it. "There is no scientific data supporting the idea that dogs perceive humans as dogs. On the contrary, beautiful research has shown that both dogs and cats form a bond with us similar to that formed by children," Dantas says.

Dogs look to us for guidance, read our body language, and get scared if their owner is suddenly causing them pain and fear. A species at the cognitive level of a human toddler, moreover, is likely to feel confused when their human family is sometimes nice to them and then sometimes yells and lashes out. "This can lead to emotional trauma," Dantas says.

Dantas, who trained for years to learn to prevent and treat fear, anxiety and stress in animal patients, is adamant that humans who see themselves as a pack leader are literally abusing their dogs, causing fear, anxiety disorders and, ultimately, the very aggression they seek to end.

She tells her clients to beware of devices like shock and choke collars sold in pet stores and even recommended and used by some dog trainers to forcefully show a canine that its human is supposedly the alpha pack leader. She also warns against alpha rolls (a discredited but still used training technique that involves pinning dogs on their backs and holding them in that position, sometimes by the throat, if the canine disobeys), jerking a dog's collar and other types of physical and emotional punishment.

"Besides hurting the bond between the human family and their best friend and predisposing dogs to further behavior problems, those techniques lead to aggressive behavior with no warning signs, which is extremely dangerous and resistant to treatment," Dantas warns.

The most efficient way to train your dog is to calmly reinforce positive behaviors. For example, if your dog shows stress or signs of aggression when on a leash (pulling, lunging, growling or barking), instead of yelling or punishing your pet, Dantas advises gently teaching him to pay attention to you so the dog disengages from whatever stressful stimulus was causing distress.

What if you've already barked up the wrong tree, believing you had to be the alpha, and dominate and shame your dog to obey? It doesn't mean you are a bad person or the situation is hopeless. But it may be a good idea to consult a veterinary behaviorist for help. To find a certified veterinary behaviorist, visit the American College of Veterinary Behaviorists' website (dacvb.org).

CHAPTER 3

Pure Genius

Many animals show their intellect in surprising ways, perceiving and experiencing the world around them to their own advantage.

The large brown bear has no more brain cells in its cortex than a cat, and 50% less than a raccoon.

The Shape of *Intelligence*

Like a scoop of jam spread on toast, a brain may have a thick carpet of neurons or a sparse supply. A neuroscientist explains how the differences translate.

Suzana Herculano-Houzel was a science communicator with a PhD working at a museum in Brazil almost 20 years ago when she discovered something strange: The general public, including 60% of college-educated people, thought that they used only 10% of their brains, "which is, of course, nonsense," she says. "We use the whole brain the whole time." Other myths were prevalent, even among scientists: For instance, the idea that the human brain had 100 billion neurons was in textbooks everywhere, but no one had actually done a count. "I realized that we didn't know the first thing about how different brains are made up," she says. So that's what Herculano-Houzel set out to do, starting with mice and rats, other mammals and then species of all sorts. Today, as a neuroscientist at Vanderbilt University, she spends her time studying brains across species—from the lowliest insects to primates like us. Her discoveries, elaborated in the Q&A below, are extraordinary: Every species has a unique, one-of-a-kind brain, and general intelligence and life span seem to depend on the number of brain cells we have.

How is it that scientists didn't know how many neurons different brains, including ours, really had?

First, out of necessity, you just trust that other people know what they are doing when they write a textbook claiming that the human brain, for instance, is made of 100 billion neurons. But when I searched for the study proving this, I realized it did not exist. When I did the count myself, I found the human brain had about 86 billion neurons. The other problem was we didn't really have a good method to give us those numbers in the past.

Where was the error?

The usual thing was to take a whole brain, slice it into very thin sections, put it under the microscope and then you would simply count the cells. It was like taking a poll using just tiny samples before an election when the real population is highly heterogeneous, so chances are you're not going to get a representative of the whole.

You solved the problem in a surprising way. Can you explain?

I thought that if the problem was the heterogeneity of how the neurons were distributed in tissue, let's literally dissolve the tissue and just collect the cells. If I could turn the brain into soup, then it should be straightforward to agitate the soup to make it homogenous, and then go back and count the nuclei of the cells. If I have this many nuclei in a given small volume I can extrapolate to the whole volume, and, long story short, it really works beautifully.

What animals did you start with, and what did you find?
We started with rats and mice, and that was important, because there were two brain structures in those animals where numbers of neurons were known: The cortex of the rat and the cerebellum of the mouse. Our numbers matched, and that proved our method could work.

After that you went to other species. What were they?
We did a number of other rodents like guinea pigs and the capybara, which is a big, dog-sized rodent that lives in Brazil. That gave us a clear-eyed notion of the proportionality between the numbers of neurons and the size of brain structures in rodents.

You also studied primates.
We studied marmosets and different types of monkeys and showed that there was a very simple proportionality. The more neurons a primate has in the cortex, the bigger its brain, and that held right up the line to humans. The primate brain was put together differently than the rodent brain.

The structural difference between rodent brain and primate brain come as a surprise to you. Why?
The assumption was that if you took any two brains from any of 2 million species, if they were the same size they should have similar numbers of neurons because the brain is a brain is a brain. If you stick to mammals, the idea was from the cat brain to the cow brain to the bear brain to the monkey brain, they should all follow the same basic rules, with the only difference being size, like a universal law. But we showed that the rodent cortex was completely different than the primate cerebral cortex. You will always find more neurons packed into the same area of a primate brain. If you compared our brain to a gigantic rodent brain of the same size, you would find humans have seven to 10 times more neurons for a given area. Primate brains follow a different set of rules.

What about the long-held association between brain size and body size?
The presumption was that bigger animals have bigger brains. So an elephant brain should have even more neurons than a human brain. But if that's true, how come we are studying elephants and not the other way around?

But you proved that wrong.
Humans have the biggest primate brain, and we have the most neurons of any species inside the cerebral cortex. Our primate cortex is the part of the brain that gives us complexity and flexibility. It's possible, based on our research, to make the argument that it is the number of information processing units—the number of neurons—that determine your capabilities and matter the most, not brain size.

Rodent smarts: The guinea pig has 44 million neurons in its brain, compared to 14 million for the house mouse.

The Brain Pyramid

Nonprimate carnivores, including cats and dogs, have neurons that scale to the size of the brains. Two exceptions: Raccoons have far more neurons than expected, and the brown bear has far fewer.

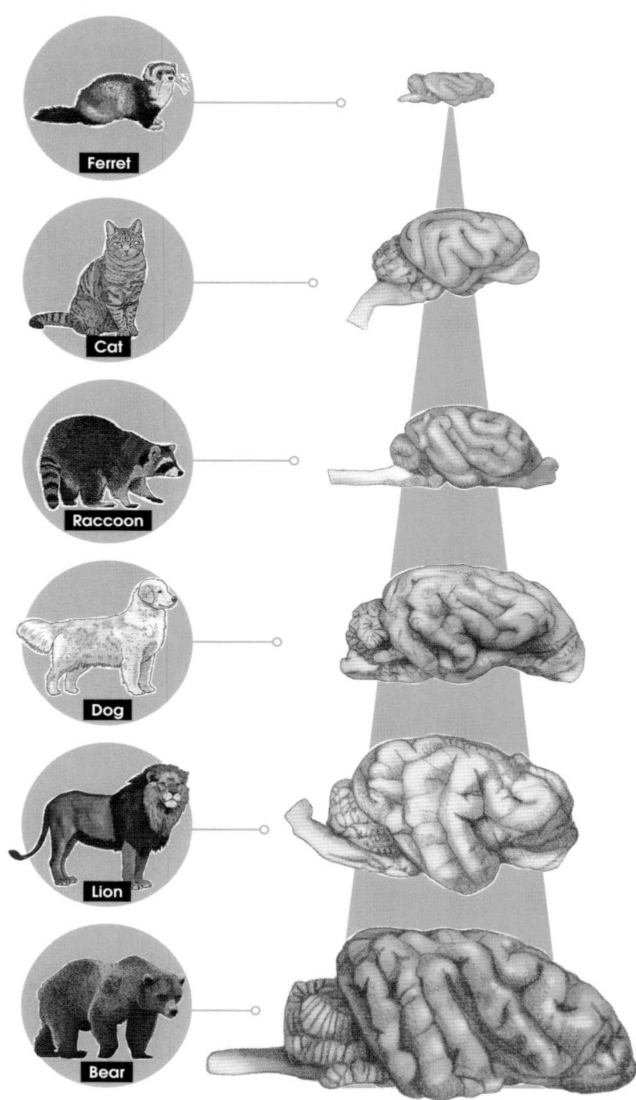

Elephant brains have most of their neurons in the cerebellum, a structure that coordinates movement.

You've studied the brains of 70 species. Is each one specialized?
Rather than calling any brain specialized or even special, the main message is that they're all doing incredibly different things, reflected by different ways of putting brains together.

Can you give an example of how this plays out? Let's talk about the crow, which is so intelligent it has been compared to chimpanzees or even human children.
The crow, in its equivalent of the cerebral cortex, has as many neurons as you find in a monkey. Yet the monkey brain is the size of your fist and the crow brain is the size of a walnut. You find something similar when you compare a monkey cortex to the cortex of a giraffe. The giraffe cortex is much, much bigger, but it only has about as many neurons as that smaller monkey cortex. They are made in different ways.

When you say different brains are made in different ways, what do you mean?
I am talking about how many neurons they have, how large the average

neuron is, and how a scoop of those neurons is spread laterally over the surface in terms of thickness. Say you have a piece of toast and you have jam. The jam can be spread on the toast to be more or less dense with larger or smaller chunks of fruit. You can take that one scoop of jam, and you can spread it very thinly over a very large surface of your toast. Or, you can spread it very thick over a small surface of toast. In mammals, you can take a similar scoop, with the same number of neurons, and spread it thick and wide or thin and narrow. In the end, the total number of neurons determine the complexity you see. Primate jam is different from rodent jam, which is different from marsupial jam and so on. There is also enormous variation in how that jam is spread from giraffes to antelopes to cows.

So how would this work in the octopus, with its eight arms?
There you have a completely different type of brain, in terms of where the neurons are and what the structures are and how the whole thing is organized. The octopus brain can be very small but has many more neurons than you would expect. To put it in range, the octopus brain is about the size of a mouse brain, but it has as many neurons as a rat brain—which is five times the size of a mouse brain.

What is the main takeaway?
Right now, what I am proposing is that what matters in terms of cognitive capability, at least when you're comparing mammalian brains and probably also bird brains, is just how many neurons you have in the cerebral cortex. How those neurons are spread doesn't matter. The size of the animal's body doesn't matter. Not even volume occupied by those neurons matter—just the number.

What is your research focus for the future?
I've found that the best predictor of a species' longevity is the number of neurons in its cortex. It takes primates, especially humans, longer to develop and longer to reach their sexual maturity. Primates live longer than other warm-blooded vertebrate species. The number of neurons correlates to life span for a huge range of animals. I would like to understand the link.

What about the Darwinian idea that all species are evolving and improving, becoming bigger-brained and more perfect—and perhaps longer-lived—as time goes on?
That's not what I see in the brains I look at. To the contrary, I find a much kinder, gentler type of biology where you don't need to be optimized, you don't need to be great, it's really just whatever works. Whether the individuals in a species tend to live one year, 10 years or 100 years, their brains all work because, obviously, the animals are all alive and well, and they do perfectly fine in biological terms of being able to eat, move and reproduce to keep the species around.

Size Isn't Everything

In some species, a tiny brain packs a ton of cognitive power.

Capybara
The world's largest rodents are intensely attached pack animals that live in the wild as bonded pairs. Capybaras have small brains, but ethologists say their emotional attachment indicates they possess a "theory of mind," in which one mind is cognizant of another, which can apply to animals.

Octopus
This brilliant invertebrate has been known to break out of an aquarium tank and find its way back to the ocean. It packs as many cells as a rat brain in a mouse-sized brain—and has a truly distributed nervous system, with more neuronal cells networked through its arms than in the brain itself.

Raven
They have among the largest brains of any avian species, though small by mammalian standards. They're so intelligent they can plan future events. A raven was observed coming upon a carcass. The next day he brought a group from his flock to retrieve the meat.

A free-range chicken stands on a ladder, which gives her access to the crafted structure she shares with other hens.

Birds atop the pecking order like to roost on the highest perch within the henhouse.

Uncooped Backyard Chickens

When we think of farm animals as food, we might forget their inner qualities: emotional, empathetic and smart.

On a windy night in April, a bear came down out of the woods, sauntered into our backyard and ripped the door straight off our chicken coop—hinges, trim, and all. I awoke the next morning to our most beloved hen, Crooky Comb, clucking persistently below our bedroom window. Perplexed as to how she got out of our latched, sealed chicken coop, I traipsed out into the gray morning to find the coop door in the yard, a dead chicken just beyond it and another partially eaten one nearby, the other five hens nowhere in sight. Tooth marks were up and down the edges of the door, angled holes where the bear had bitten right through the wood.

I stood in disbelief. We'd anticipated predators small and wily, not those who would opt for brute force. Nighttime renders chickens utterly vulnerable and therefore inert. A bird who enters the coop after dark will not be able to make her way up the rungs to the highest perch. Caught outside after dark, she will bed down where she is. Darkness signals stillness—and without the protection of our coop, our chickens are sitting ducks out in the open night.

I have, since that night, imagined my way into my hens' experience. I sometimes picture myself as a chicken perched on the highest rungs of the coop, where our birds on top of the pecking order liked to roost. One moment it is dark and quiet, and I'm nestled against my flock mates. The next, I'm awakened by the sound of the bear scratching at the coop, listening to its loud snuffling at the cracks in the door. A shiver of terror runs through me as I hear the claws scrape at the hinges, wrenching them off. I imagine bear breath: tons of saliva festered in a closed maw—morning breath after a five months of sleep. What might have been running through my mind as my friends, my sisters, around me were plucked from their perches and eaten alive?

College science courses warned me about the dangers of anthropomorphism. But now that I am a chicken keeper, I can't help it. My own imagination seems crucial to understanding my birds' well-being, and I wonder if too many years of separating ourselves from the minds of animals has enabled objectification—a mindset that allows for factory farming and all the questionable ethics surrounding treatment of livestock today.

A large-scale farmer who depends on selling chickens must think of chickens as a commodity. It's dangerous for such a farmer—at least, one who wants to put food on her table and a roof over her head—to consider the moral and ethical implications of chicken individuality while she strives for profitable production and economies of scale.

It's been convenient, therefore, to just say that chickens are dumb, not much more than mindless squawkers—just like we insist that lobsters can't feel pain (an idea that has been scientifically debunked); that the cow never knows what hit it (it can see it coming, just like you can); or that the pig actually likes wallowing in its own feces (they'll be tidy if they have the space to be).

Animals that we deem "intelligent"—such as horses, dogs and dolphins—are esteemed; and in the United States, we disdain the thought of eating them. It may be true that those animals are intelligent, but we have vastly underplayed the intelligence of the animals we do eat, including chickens.

I reached out to neuroscientist and animal behavior expert Lori Marino, to get a sense of why. "Acknowledging who chickens are—bright birds who have lives to lead and do not want to be killed—is inconvenient for our gustatory pleasure," she claims. "Their lives matter to them."

Senior scientist for The Someone Project (a joint venture between the Kimmela Center for Animal Advocacy and Farm Sanctuary), Marino has written several scientific articles about chicken intelligence. "Based on our review of the research done with chickens, it's clear they are cognitively, emotionally and socially as complex as many other birds and some mammals. They are intelligent, reasoning and aware beings. They care about each other and how they are treated."

Did You Know?
Chickens have feelings like anxiety and fear. They have empathy for other chickens and express unique personalities: Some are introverts; others, queens of the roost. They anticipate events and can even do basic math.

For example, chickens are capable of a kind of deductive reasoning known as transitive inference. They understand quantity and can perform basic arithmetic. They can estimate time intervals and anticipate events. And they have complex forms of communication, as any chicken keeper knows.

I can tell my hens' alarm calls from the egg-laying ones or aggressive hen-pecking calls. They have different ways of behaving too: When they all become silent and stand stock-still, it typically doesn't take long for me to locate the curious raptor perched nearby. When a dog runs by, they'll scatter, emitting high-pitched, distressed calls.

They even have self-control. Given the choice between waiting two seconds for

How to Keep Your Chickens Happy

Chickens are happiest when they are given a setting that replicates the natural home of the wild jungle fowl—the "wild counterpart," as neuroscientist Lori Marino calls them, to our domesticated chicken. An appropriate setting for chickens is "one that affords all the behavioral opportunities the chicken is capable of." Key elements for an environment for chickens, Marino claims, include: • Sky over their heads • A place to perch • A soil substrate that drains easily and allows chickens to scratch • Things in their environment that they can explore and manipulate to satisfy their social and intellectual needs • Protection from predators

Chickens feel each other's pain and have complex forms of communication.

Chickens can recognize and distinguish among more than 100 faces of their own species.

The whole family can bond around the care and feeding of backyard chickens, which provides a sense of fun and responsibility—as well as organic eggs.

some food or six seconds to hit a food jackpot, the chickens overwhelmingly held out, demonstrating an ability to actively optimize future outcomes.

Chickens have feelings—anxiety, fear. They have empathy too, and they will become emotionally distressed when observing another chicken experiencing pain or stress. They have personality: Crooky is our affable companion; Welsie is our goofy chatterbox; Dotty's the recalcitrant observer who prefers a freshly picked dandelion leaf over the squirming carapace of a potato bug. Chickens have self-control, self-awareness and self-interest—which is to say, they recognize themselves as selves, as complicated beings with wants and needs and preferences.

I got chickens for the eggs, much like millions of other Americans over the past decade. Backyard chicken-keeping is on the rise. The most recent comprehensive report regarding the number of backyard chicken keepers in the U.S. came out in 2013. It found that, while less than 1% of households in four major American cities kept backyard chickens, 4% planned to have them in five years. In my state, Massachusetts, Michael Cahill, director of the Division of Animal Health for the Massachusetts Department of Agricultural Resources, reports that the number of locations having small chicken flocks (one to 11 birds), nearly doubled (going from 870 locations to 1,725 locations) between 2005 and 2013. And the trend is growing. Everyone knows someone who's got a backyard flock, it seems.

Raising chickens for eggs felt more ethical to me than buying store-bought eggs. I have a decent-size yard. I've cared for chickens before. But now that I have them, I wonder about my assumption. Can I keep them safe enough? Do I know enough about their needs—and desires? Having chickens under my watch has caused me to wonder if keeping chickens for eggs is still exploitative, even if I attempt to give them a safe and comfortable life. I posed the question of ethics to Marino, and her response was that she does see backyard chicken-keeping as ethical but encourages those interested in doing so to adopt from shelters and rescue groups rather than buying from commercial hatcheries, which, she claims, "maintain cruel practices." She says, "If someone wants to keep backyard chickens just for the purpose of getting free eggs, then they need to think twice about whether they will have the dedication and love needed to care for these creatures on their own terms. Yes, love. Just like all other animals, chickens need to be cared for and to feel cared for."

Chickens have self-control, self-awareness and self-interest. They recognize themselves as selves.

Later in the morning after the bear attack, my husband found another of our hens perched low in a spruce tree; that afternoon, another came wandering back. They acted traumatized, he said. The next day, I found the remains of one of our Wyandottes up in the woods, about a mile away. The other two had disappeared without a trace, except for a few stray feathers that still linger in our yard. Were they to have escaped the bear's maelstrom, they wouldn't have made it long, with our hungry neighborhood foxes and hawks.

I would say I loved our chickens before the bear struck. But since then, I've been paying particular attention. Perhaps it's easier with just three birds rather than eight. I knew their personalities before, but now they are three distinct individuals, with different fears, desires and ways of being in the world. I would go even further to say I've forged a particular kinship with Crooky Comb. My 1-year-old daughter and I will sometimes commune with our chickens for hours at a time. We will wander around their yard, turning over stumps and rocks, and they'll trail closely behind us, scratching for ants and beetles. They'll peck at flecks on our shoes, making my daughter laugh. Crooky will come especially close, and I've taught my daughter how to reach out her hand, a signal that makes Crooky hunch down to the ground. Science tells me that it's an act of submission, a sign she knows she's subordinate. But something else tells me that there's autonomy in her choice: the other birds don't let us pet them this way, for example. I respect their wariness. Crooky makes little clucking noises as I stroke her gold feathers. And sometimes, even as I lift my hand away, she'll stay squatted down next to me, as though she just wants to be close by. *—Amanda Giracca*

Amanda Giracca is a lecturer in SUNY Albany's Writing and Critical Inquiry program and a contributing editor to Vela *magazine.*

The Perils of the Rooster

The male of the species is harder to handle, so be prepared for rowdy behavior.

If you mail-order a run of chicks from a commercial hatchery that claims it sexes day-old chicks (meaning, it promises to send you hens only), be prepared, given the high margin of error, to receive a rooster or two. Roosters can be noisy and aggressive toward adults and children—and sometimes other chickens. If you are prepared to keep your rooster, check your city ordinances to see if that's allowed. If you end up with an accidental rooster that you cannot keep, you may be able to locate a nearby rooster-rescue sanctuary that would be willing to take the bird off your hands. If you are considering a backyard flock, have a plan for what you'll do if a rooster arrives.

Some humans have reported feeling the buzz of dolphin echolocation when in the water with them.

Strange *Sensors*

Nonhuman species wield a host of skills to perceive and navigate the world.

Living in our human world, we use five basic senses—including touch, sight, hearing, smell and taste—plus a few more subtle ones, such as understanding how our bodies are positioned in space (proprioception) and the ability to detect force (mechanosensation). Researchers have long been studying these perceptions that some like to call the sixth sense. But as it turns out, our nonhuman neighbors possess a range of senses that we can't fathom. Here, we take a closer look at the unique ways that some animals experience and interact with the world. Included are dolphins that navigate deep underwater using echolocation, snakes and bats that use infrared heat vision to zero in on their prey, octopuses that see with their skin and bees that find home by sensing the Earth's magnetic field. Read on for more of their mysterious skills.

DOLPHINS
FINDING THEIR WAY THROUGH ECHOES

How do odontocetes, or toothed whales (a designation including dolphins and porpoises), swim deep underwater where light is minuscule without crashing? Echolocation is the process by which toothed whales essentially see without eyes by emitting sounds that travel through their foreheads and bounce off objects. The echo travels back through the dolphin's lower jawline and into its eardrum, painting a detailed portrait of the landscape, including obstructions or food sources. How detailed is that portrait? A toothed whale can distinguish between objects of less than half an inch in diameter from 50 feet away, making echolocation a far more reliable nighttime navigational tool than sight. But scientists still don't know whether toothed whales emit their echolocation sounds from their larynx or nasal plug or how exactly their heads are able to focus this noise.

PYTHONS, BOAS AND PIT VIPERS
FEELING INFRARED HEAT

These reptiles are gifted at catching prey at night, but they don't use traditional vision to track them down. Instead, pythons, boas and pit vipers have holes on their faces called pit organs that detect infrared heat from any warm body within a 1-meter range, letting snakes feel their prey as they slither by. The pit organ is part of the snake's somatosensory system, which detects touch, temperature and pain. Boas and pythons have three small pit organs. Pit vipers have two larger, more complex pit organs containing a suspended

The southern ridge-nosed rattlesnake has a heat-sensing pit on each side of its face to detect prey.

Snakes sense pain, touch and temperature but not photons of light.

sensory membrane inside a hollow bony chamber, maximizing the detected temperature differences between warm-blooded animals they view as food and everything else. Infrared radiation heats up the pit organ's membrane tissue and sounds the alarm at about 82 degrees Fahrenheit (roughly the temperature a snake would "feel" from a mouse a meter away).

OCTOPUSES
SEEING WITH SKIN

Octopuses are renowned camouflage artists, changing their skin color to blend in with the environment, avoid predators and communicate with fellow octopuses. How does it work? Embedded right under their skin by the thousands are specialized cells called chromatophores that contain pigments. The pigmented chromatophores are surrounded by muscles that expand or contract, making the pigment more or less visible and thus altering the color of the octopus as a whole in response to environment. For years researchers thought this was a passive response, allowing the colorblind octopus to respond to threats without any top-level involvement of the brain. But now, it turns out, we understand the chromatophores may function like eyes. After removing the chromatophores and studying them in the lab, evolutionary biologists Desmond Ramirez and Todd Oakley of the University of California, Santa Barbara, revealed they respond to the same wavelength of light as pigments in eyes called opsins, suggesting the presence of interactive light sensors in the skin. In eyes, opsins send an image to the retina and then the brain. In the octopus, more loosely structured opsins may communicate brightness, showing why it can so quickly hide from predators. Further studies show these cells also respond to touch.

GECKOS
DETECTING COLOR AT NIGHT

When it's night outside with barely any moonlight, human beings can make out only dim, shadowy outlines. Geckos don't have that problem. They can see colors in the dark—thanks to a process called photoreceptor transmutation, hypothesized by researchers more than 65 years ago but not backed by evidence until now. Basically, the gecko's photoreceptors, once composed of retinal cones (which allow day vision and the ability to see color), gradually

evolved over time into retinal rods (which allow shadowy night vision). But these retinal rods have somehow retained both rod- and conelike properties, similar to organs found in some snakes, according to a study in the journal *Evolution*. When the gecko shifted from daytime to nighttime patterns, its all-cone retinas evolved over generations into all-rod retinas that retained the ability to discern colors at night—a new superpower to help it survive in the dark.

HONEYBEES
MAGNETIC SENSE

How do honeybees find their way home? These social insects often travel great distances to forage. In addition to visual cues gathered midflight, honeybees rely on magnetoreception, or the ability to detect magnetic fields, to return to their hive. The exact source of their incredible magnetoreception power isn't known, but it appears that iron granules in their abdomens may increase their sensitivity to the Earth's magnetic field. The granules undergo changes in size and orientation when exposed to a magnetic field, giving honeybees immediate navigational feedback. Honeybees use this unusual sense to orient themselves and locate familiar places. They even build their comb in the same magnetic direction as the parent hive. Honeybees are influenced by magnetic fields even in the dark, when they are unable to gather any other directional clues.

VAMPIRE BATS
INFRARED VEIN DETECTORS

The vampire bat is the only mammal known to survive entirely on a blood diet—and it needs to feed on a few tablespoons daily. But how does it know where to strike to find a vein? The vampire bat has developed infrared sensors on its lips especially evolved to locate blood vessels in its prey. These sensors are an adaptation of the same genetic sensor that alerts human beings to dangerously hot stimuli, like scalding tea or an oven on high. They can detect regular body temperatures and seek out the best veins from up to almost 8 inches away.

ELECTRIC EELS
LIGHTNING RODS

The electric eel makes for a surprisingly sophisticated predator, relying on special cells called electrocytes to store energy and discharge it in the form of electrical fields. Not only do these electric fields offer protection, but electric eels can also use them to locate and immobilize their prey. Once the hunt is on, electric eels can generate electric pulses as strong as 500 hertz, which they employ to remotely control their prey. A high-voltage pulse can render prey immobile or cause them to twitch involuntarily, making them easier to locate and attack. Some electric eels can even leap out of the water to electrify their target.

Electric eels can reach up to 8 feet in length and weigh as much as 44 pounds.

Rats can remember the location of scented items in a maze.

A rat in a maze is not just functioning on instinct or senses. It can literally remember the past.

Total *Recall*

Do animals live in a perpetual now, or can they remember the narrative arc of their lives?

My two cats and I have a morning ritual. They tap my face to wake me up, then about 20 minutes later, I unfurl their bag of food and they come running, eager for breakfast. In the evening, the same sound of the food bag crackling open sends them darting back into the kitchen for dinner.

At some point they learned that a rustling bag means chowtime, and now they remember this cue. But can animals recall more complex events and experiences, the way people are able to? For years, the question was heretical—experts held that animals lived largely in the present, recalling cues essential for survival though not much more. But a host of surprising studies in recent years has researchers increasingly convinced that some animals can, in fact, recall past events and retain real memories of their lives.

VARIETIES OF MEMORY

There's more than one type of memory. Semantic memory—remembering something that was learned—is what my cats are using when they respond to my dinner cue. We use semantic memory when recalling that the capital of Texas is Austin, for example, or retrieving facts for an exam.

The second—and more complex—kind of memory involves consciously going back to previous events and replaying them in the mind. We can remember what we did last Saturday afternoon or when we last took out the recycling. This more sophisticated kind of memory, which requires recalling past episodes in our lives, is called episodic memory.

These complex memories are challenging to test for in animals, primarily because researchers can't just ask them what they did at 2 p.m. last Tuesday. So scientists have developed a set of criteria for episodic memory in animals. First, the animal has to show that it's aware it's remembering something, says Victoria Templer, a behavioral neuroscientist at Providence College in Rhode Island who has studied memory in monkeys and rats. Then, once this awareness is confirmed, memories must exhibit at least one of the "three Ws": what, where and when. And the animal has to show that it remembers an event or experience in the order that it happened.

Many animals have passed the first test of episodic memory, the retrieval, Templer says. For her own research on

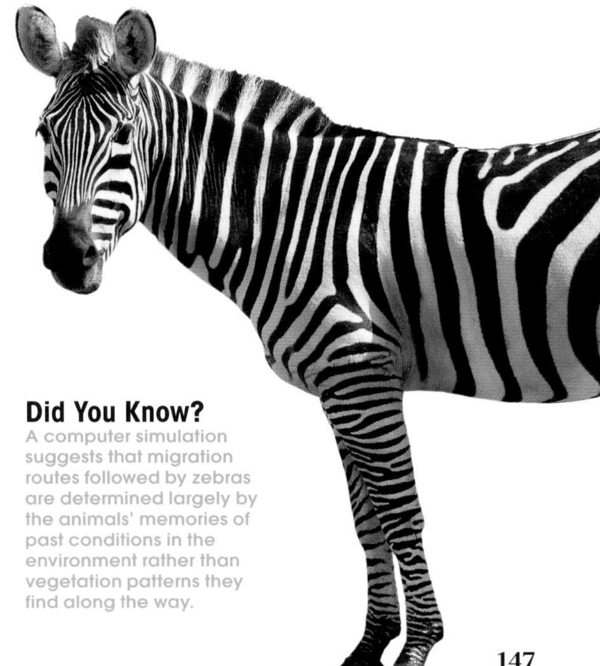

Did You Know?

A computer simulation suggests that migration routes followed by zebras are determined largely by the animals' memories of past conditions in the environment rather than vegetation patterns they find along the way.

this question, Templer and her team trained rhesus monkeys to knock over cups to communicate whether they remembered where a hidden cache of their favorite food was located. If they declined to knock over any cups, they'd still receive some food, but not the precise kind that they preferred. The team found that "they're more likely to bail out of a test and get a less preferred but a guaranteed reward" if they couldn't remember where the good stuff was, Templer says.

Cats are aware that the rustling of a bag means someone is getting the food, and chow will soon be served.

Narrative life memory was considered a hallmark of humanity. But now, we realize, other creatures can do this too.

Determining whether animals can recall what an event was and where and when it happened—the three Ws—is more complicated, but researchers cite several compelling examples of at least one of these features. A seminal 1998 study in *Nature* looked at western scrub jays, a raucous bird in the corvid family—a group known for its cognitive abilities. Researchers there found that scrub jays can remember where and when a food cache was stored and prioritize which one to use first based on how long it will last. Given the option of two caches, one of perishable wax worms and the other of nonperishable peanuts, the jays went for the worms first.

A DOG'S LIFE

It makes sense that animals can remember important things like where food is and how soon it needs to be eaten. But a 2016 study published in *Current Biology* found that dogs, at least, are able to recall even nonessential memories after a period of time.

To test this, the researchers had the owners of 17 dogs of various breeds train their pups to mimic six interactions with different types of objects, including an umbrella, a chair and a bucket. That created the expectation that when their owner executed an action involving one of the objects, the dogs were to do the same thing.

But in the second part of the experiment, the researchers had the owners spring a surprise on the dogs. This time, after watching their owner interact with one of the objects, the dogs were told to lie down instead of mimicking them. The next time the owners played with an object, the dogs immediately lay down again, expecting the same drill. But instead, the owners paused for one minute and then commanded the dogs to mimic their trick with the object. So far, so good.

Then in a final step, the owners went through the routine again, this time extending the delay between demonstrating the trick and issuing the command in an hour. The researchers wanted to see if even after being distracted the dogs could still remember their owners' behavior and copy it.

Not surprisingly, almost all the dogs mimicked their owner when immediately commanded to. But even when the pups had to remember after distractions and delays, they had a decent success rate: 58.8% were able to mimic their owner after the minute-long delay, and even after an hour, about 35% of the dogs remembered what to do.

Dogs are not alone. An earlier study showed that dolphins could remember the calls of other dolphins they had associated with, even decades later. Apes appear capable of remembering

Scrub jays can use episodic memory to strategize for survival.

This western scrub jay is so advanced it can remember where and when a cache of food was stored.

Dogs have the ability to remember what their owners did hours before.

events that occurred several years in the past as well. (See sidebar, right.)

RODENT REPLAY

Animals like apes, dolphins and jays are well-known for their cognitive sophistication. But now there's mounting evidence that even rodents, viewed as unintelligent vermin in many cultures, may have episodic memories as well. Research published May 2018 by Indiana University neuroscientist

Chimpanzees and orangutans can recall the location of tools that have been stored away years before.

Jonathan Crystal's lab found what he believes is the first conclusive evidence of an animal mentally replaying past events when recalling them—the holy grail in episodic memory research. He knew from a previous study he conducted that rats could remember at least 30 different events, but could they travel back in time and relive the events in their minds, the way we can?

To find out, Crystal and his team trained 13 rats to visit a series of scented items in a mazelike arena in a particular order, then increased the delay time between the presentation of items at various points in the sequence to ensure the rats were relying on episodic rather than short-term memory. The rats successfully completed the test in the right order about 87% of the time. "We think that this doubling of the delay rules out all of the nonepisodic memory explanations," Crystal says.

To make sure, though, the team also temporarily suppressed activity in the hippocampus—the site of episodic memory in the brain—and ran the test again. Sure enough, the rats performed poorly this time, confirming that earlier they had been using the hippocampus—and therefore episodic memory.

While none of these studies tick all the boxes for humanlike episodic memory, collectively they challenge the belief that humans alone are capable of this kind of recall, many researchers say. "I think there's pretty strong evidence that animals do have various aspects of episodic memory," Templer says.

As the evidence for episodic memory in animals mounts, it forces us to reconsider our relationship to them. Episodic memory has long been one of the primary cognitive abilities that we presumed set humans apart from the rest of the animal kingdom. If it turns out they, too, are replaying the past in their minds, living much richer and more complex lives than we thought, should we treat them differently?

We may have come a long way from René Descartes' 17th-century view that animals are merely automatons, undeserving of compassion, but it remains to be seen whether we're ready to accept them as more like us.

—April Reese

April Reese is an independent environment & science journalist.

Triggered Memories

A cue helps orangutans and chimps remember the location of long-hidden tools.

In *Remembrance of Things Past*, French writer Marcel Proust describes how the taste of a madeleine cake brings childhood memories rushing back. This capacity to trigger scenes from the past has been considered distinctly human. But a Danish research team has shown that chimps and orangutans—the two most cognitively sophisticated primates aside from humans—can be triggered to recall the past. In exchange for a food reward, chimps and orangutans deftly found tools hidden in a remote room the researchers had revealed to them either four years or three weeks before. A control group of animals could not find the tools, showing that success resulted from triggered memory, not happenstance. The results suggest that our autobiographical memory might have evolved millions of years ago to aid problem-solving—and triggering elements could help pull up useful memories when necessary.

From schnauzers to spaniels, teacups to giants, dogs of every size, shape and breed can help make the world a better place.

Canine protectors save wildlife, aid the blind and provide therapy for autistic kids.

Dogs Save the Planet

Pups are at the front line of conservation efforts to protect the endangered species of Earth.

Pepsi leans over the bow and scans the water off the coast of the San Juan Islands. She is cruising along in a small motorboat with her research team, looking for orca feces or, as the scientists call it, scat—floating masses with the consistency of pancake batter, often less than a tablespoon in volume, stinking of rotten fish and ranging in color from white to green. It's quite a challenge to spot the scat with the reflection of the sun on the waves of Bellingham Bay. Even more challenging is the race against time. If Pepsi doesn't find the scat within roughly half an hour after deposit, it sinks and is lost. When she does detect scat, she directs her colleagues across the water so they can collect the samples.

But Pepsi isn't using her eyes. And she isn't a person. Pepsi is a conservation canine. Also known as wildlife-detection dogs, conservation canines are trained to sniff out threatened and endangered species or other wildlife, from orca to owls, foxes to frogs, bats to bobcats.

Pepsi's work assists scientists trying to find the causes of orca decline in the Pacific Northwest. Their scat can be analyzed to give insights into their diet and any toxicants present in their bodies. Other ongoing projects with canines working alongside conservation scientists include researchers from the Desert Research Institute, tracking threatened desert tortoise in California; a team at Dalhousie University, searching for ribbon snake in Nova Scotia; and a consortium of agencies, mapping rare salamander distribution in New Mexico.

With 220 million scent receptors, a dog's olfactory sense is at least 10,000 times greater than a human's.

Highly trained dogs are relatively recent additions to the science of conservation biology. In the late 18th century, researchers were using direct observation and specimen collection to study wildlife. This then gave way to tagging, affixing some kind of visual marker, which proved helpful for tracking individuals.

"Then you could know not just that a flock of Canada geese arrived at a particular pond every year; you could know that the same flock arrived," says wildlife biologist Kristoffer Whitney at the University of Wisconsin-Madison. The ability to follow specific flocks of migratory birds year after year led to such developments as the flyway concept—the notion that several routes on the planet are used by many species of birds.

Over the past several decades, researchers have been tracking individual animals with radio collars and radiotelemetry tags. Even more recently, conservation biologists have started following wildlife with unmanned drones.

MASTERS OF SCENT

It wasn't until 1997 that Samuel Wasser, director of the University of Washington Center for Conservation Biology, started collaborating with trainers who were teaching dogs to find narcotics by scent. Instead, Wasser found, he could train the dogs to focus their noses on wildlife. There's good reason to trust dogs with this task. A human's sense of smell is isolated to a small part of the roof of the nasal cavity, but dogs have a dedicated area in the back of the nose for olfaction. It's filled with turbinates, tiny bony structures that sort odor molecules based on their chemical properties. Additionally, dogs have a second scenting structure that humans do not: Jacobson's organ, also known as the vomeronasal organ, which is located in the bottom of dogs' nasal passages. This organ is able to detect pheromones, those behavior-altering chemicals animals release into the environment, allowing dogs to identify specific species. With an average of 220 million scent receptors, a dog's olfactory sense is at least 10,000 times more acute than a human's.

Without the help of the dogs, humans or their cameras might spot only the very largest scat samples from the boat or sky. Dogs allow researchers to include samples that would have previously gone undetected. Moreover, scat gives insight into much more than just population counts—including factors such as diet, physiological health and environmental stress.

Despite the importance of scent, wildlife-detection dogs are not the canines with the strongest sense of smell. Bloodhounds are thought to be the best scenting breed on the planet, with 300 million scent receptors so reliable that bloodhound-based evidence is admitted as court testimony. Yet the dogs selected to be conservation canines are often Labradors. While they have a wide range of scenting ability, "Labs are one of the most behaviorally malleable breeds," says James Tantillo, a lecturer at Cornell University, executive director of Orion the Hunter's Institute and expert on animal ethics and the philosophy of hunting. He acknowledges that English setters and pointers have better noses but adds, "Labs are less risky behaviorally, less likely than other breeds to fight or bite."

A dog's nose is designed to filter out odor particles from those of air, contributing to its highly attuned sense of smell.

PLAYING THE GAME

"It's all about the game for them," says Patricia McConnell, an expert on dog behavior and professor of human-animal relationships at the University of Wisconsin-Madison. "Usually dogs are selected for loving play, not because they're obsessed with smell. Play is such a great motivator." That drive is critical for dogs like Pepsi, who spend hours on the open water searching for whale scat. This kind of grueling fieldwork requires that workers, including the dogs, not give up until the task is complete.

Even with malleable personalities and strong scenting abilities, Pepsi and other dogs require extensive training. Conditioning dogs to properly detect wildlife and their scat takes dedicated handlers and a long time. First, a specific scent must be isolated to train the dogs to recognize the correct target. It's imperative that the scent be very specific. Cat Warren, author of *What the Dog Knows: Scent, Science, and the Amazing Way Dogs Perceive the World* and an associate professor at North Carolina State University, stresses the importance of not inadvertently training them on another scent. "If it's a live tortoise you want the dog to find,

This search-and-rescue bloodhound is on duty at the Red Cross.

Bloodhounds are the best scenting breed on the planet, with 300 million scent receptors.

A patient with multiple sclerosis poses with her service dog, Giles, who was trained by Canine Partners for Life in Pennsylvania.

Recipients and service dogs go through three weeks of in-tandem training.

make sure you are training dogs on live tortoise scent, not their shells or their poop," she says. "So one scientist used swabs of the necks of live tortoises to get the dogs going on the right scent."

Once dogs are cued onto the correct target, the next step is to sync communication between the dog and its handler. The conservation canines are only successful in tandem with their human counterparts. It doesn't matter that Pepsi can smell orca scat if she can't effectively tell her handler how far and in which direction to go. Each dog is a unique individual with its own cues, so much of the training therefore focuses on training the handler, not the dog, in order to read them. One dog might tug the leash, another might bark. In a training setting, with a target scent in a known location, the handlers watch how the dogs react and home in from different distances and directions. They also pay attention for the cues dogs give when they "alert," or indicate that the target has been reached. These behaviors are the ones trainers hope to reinforce and reliably replicate in research settings.

This is where play factors in. The handler rewards the dog with her ball every time she alerts to the correct target. However, it's critical to make sure that the reward is given out to align with exactly the right target. "Just because a dog alerts on a hole in the ground, that doesn't mean there's a black-footed ferret down the hole," says Warren. "If you reward the dog for finding the hole, the dog might go on and just alert on holes." Finally, the environmental conditions must be

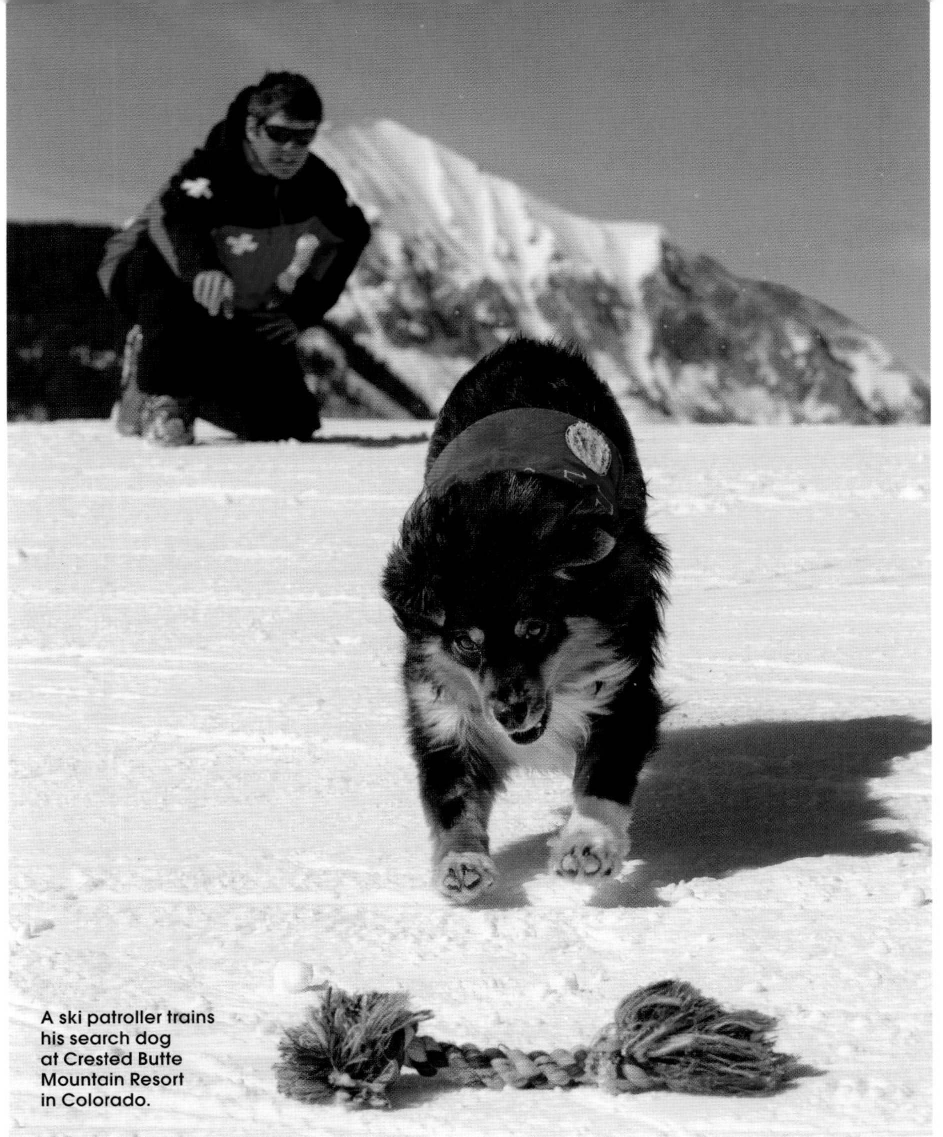

A ski patroller trains his search dog at Crested Butte Mountain Resort in Colorado.

right. If the wind is too strong or the boat isn't positioned perpendicular to the prevailing wind direction, the dog might not be able to perform.

THE DOG-HUMAN BOND

The true key to success here is the canine-human bond. Dogs have been bred for thousands of years to partner with hunters. "This is just the latest iteration of that," McConnell says.

Dogs and their handlers form a remarkable bond, each relying on the other to successfully complete their mission. In her book *When Species Meet*, science-studies scholar Donna Haraway describes the relationship between dogs and their trainers as the epitome of human-animal connection,

How "Bad" Dogs Fight Poaching

Dogs that are too energetic to serve as therapists, with no ability to quit, are ideal for catching poachers.

Megan Parker, co-founder at Working Dogs for Conservation (WD4C) in Montana, scours animal shelters looking for unadoptable, hard-to-handle dogs. "You can train a dog that doesn't stop to bring you things," she said. "It's an ideal trait for anti-poaching dogs." Take Ruger, the first anti-poaching dog in Zambia. He lives next to South Luangwa National Park, where animals are poached, snared and trafficked, and is responsible for finding elephant ivory, rhino horns and other wildlife contraband.

At first, the Lab retriever/German shepherd mix bit those who tried to work with him. But his drive convinced Parker to keep training him. She paired Ruger with a Zambian law-enforcement unit operated by the South Luangwa Conservation Society and the Zambia Wildlife Authority. On his first day, Ruger went on a job to search vehicles possibly carrying illegal goods. "It takes humans hours to search a car," said Pete Coppolillo, executive director at WD4C, "whereas it takes dogs three to four minutes." Ruger stared at a passing vehicle. The scouts found nothing. Ruger didn't avert his gaze. Inside a piece of luggage was a matchbox wrapped in plastic containing a primer cap, which ignites gunpowder in illegal muzzle loaders used for poaching. To date, Ruger has put hundreds of poachers out of business.

characterized by shared understanding, purpose and concern. This can be seen in the bond between guide dogs and their owners. "A guide-dog handler must literally trust their dog with their life. This is not hyperbole," says Andrew Hasley, a genetics researcher at the University of Wisconsin-Madison and owner of Fletcher, a guide dog he received from Guiding Eyes for the Blind more than 10 years ago. "I'm not always giving the commands. Sometimes Fletcher is telling me what to do. If a handler doesn't completely trust their dog, the relationship won't work."

It's also evident with hunters and their dogs. Brandon Barton, an ecology postdoctoral researcher at the University of Wisconsin-Madison, has been hunting with friends' dogs for more than 20 years. "The dogs are instrumental in the hunt," he says. "Houndsmen say they can tell what the bear is doing based on the sound of their dogs. Is it running? Is it in a tree? Is it fighting back? They can tell!" Tantillo concurs, calling his hunting dogs "friends for life. Sometimes I'm hunting for the dog, not even for myself. It's a unique relationship."

FORAY INTO SCIENCE

As dogs step into science, one complication has become clear: Scientists must now design studies so that the dogs remain safe. As Makie Matsumoto-Hervol, a research intern working on the same boat as Pepsi this past summer, puts it, "The dogs become your co-workers. Each one has a distinct personality and preferences

A Faithful Companion

A service dog proves unfailingly loyal to a former president.

When a Bush family spokesperson tweeted a photo of Sully, former President George H.W. Bush's service dog, guarding his owner's casket as Bush lay in state in the Capitol Rotunda in 2018, the faithful pooch captured the nation's heart. It also shone a spotlight on yet another way service dogs provide essential assistance to their human companions—in this case, keeping watch over the elderly.

Sully, a Labrador retriever, was named after hero pilot Chesley "Sully" Sullenberger III, who famously landed a plane on the Hudson River in 2009. Canine Sully first came to live with Bush in June 2018, when the former POTUS was still heartbroken over the death of Barbara, his wife of 73 years, two months before. He was also suffering from low blood pressure, Parkinson's disease and other health problems and was using a wheelchair. His doctors at Walter Reed National Military Medical Center recommended that the Bush family contact America's VetDogs, a nonprofit organization that trains service dogs, to find a pooch that would be a perfect fit for Bush 41's medical needs—as well as become a much-needed companion.

Sully arrived at the Bush family compound in Kennebunkport, Maine, accompanied by staff from America's VetDogs. He was escorted by another special guest: Bush's close friend and fellow former POTUS Bill Clinton. Sully immediately began assisting Bush with essential tasks, including opening doors, picking up items and summoning help when needed. Perhaps his most important duty: comforting the former president by resting his head on his lap.

Bush and Sully made their second-to-last public appearance together on Nov. 6, when Bush cast his vote in the midterm elections.

After Bush's funeral, Sully began a new mission, joining Walter Reed Medical Center's dog program, where he'll be helping other vets and active military personnel. During his brief time in the public eye, Sully managed to transcend partisan politics and win America's heart.

What a good boy, indeed!

Dogs and handlers form a remarkable bond, each relying on the other to complete their mission.

that you have to work with. There is this natural inclination to shower them with carrots or pet them every time you see them." Whitney expects that while this will somewhat complicate study design, it will also make it richer.

"A lot of studies of animals require you to take on the perspective of the animal you're studying in order to understand its behavior. So if you start using a dog as a research companion, you're going to begin seeing the study organism as a dog would—which could open up new insights and questions. Understanding a whale through smell is a whole new world for marine biologists."

Distinct from the vision-based technologies such as trail cameras or infrared sensors on drones, the use of scent to find hard-to-see excrement enables researchers a more unbiased sampling method. For example, without the help of the dogs, humans or cameras might spot only the very largest scat samples. Working with dogs increases the number and diversity of scat samples, because researchers are able to include samples that would have previously gone undetected.

Wildlife-detection dogs are true partners in science, qualitatively and quantitatively changing the research process. With dogs guiding scientists, literally and metaphorically, into new waters, they are emerging as collaborators in conservation.

Did You Know?

The tiny Chihuahua can make an ideal service dog: She becomes attached to one person, is very intelligent and easily fits into tiny apartments and other small spaces. This alert animal can follow not just verbal commands but also signals to provide aid.

Parakeets have large spoken vocabularies; the world record is 1,728 words.

The budgerigar, also known as the common parakeet, is a small green parrot with a short, flat tail. Budgies mimic human speech and live as mating pairs.

Birdland

A leading expert on avian intelligence discusses her extraordinary research on the brilliant minds that reside within brains no bigger than a shelled walnut.

As a 4-year-old child, Irene M. Pepperberg received a parakeet as a companion. Lacking interactions with other children and with parents who were both physically and emotionally unavailable, Pepperberg jokes that she imprinted on the tiny bird because it was the only thing that interacted with her through the day. Her earliest avian companion was the first of many who populated her childhood and created a powerful lifelong relationship she has shared with birds. Currently a research associate in the department of psychology at Harvard University and author of the *New York Times* bestseller *Alex & Me*, Pepperberg grew up to become a scientific pioneer in the study of animal minds. Her studies on the communicative and cognitive abilities of grey parrots are landmarks in the field.

Can you describe the parakeets you had as a kid? Did they talk?
Just little parakeet phrases: "Come here. Give me a kiss." "I love you." "Pretty birdie." No other children were around. My father worked full time as a teacher, was studying for his master's degree and was taking care of his mother, who was very ill. He'd kiss me good morning, and sometimes I wouldn't see him till the next morning. My mother wanted a career. She did not want to be a mother. She met my physical needs but none of my emotional needs. So my dad bought this little parakeet to keep me company.

You went from those early experiences to getting a doctorate in theoretical chemistry to working with birds. How did that happen?
I watched a *Nova* program on signing chimps, a big breakthrough in communication with nonhumans. Also a program on dolphin intelligence. Then an entire program explained how birds learn their songs and how they use them in different contexts. That's when it occurred to me—a parrot can talk. If you want to communicate with all these other animals, why not use a bird that could talk? I spent lots of time in libraries, learning what little was known about parrot intelligence and abilities. That's how I chose the grey.

Because of its intelligence or its speech capabilities?
Both. There were papers from zoologist Otto Koehler's lab in Germany showing that greys had a sense of number. Other studies showed that if you hid something behind a barrier, they immediately knew to go behind it. And pet books claimed greys were the clearest of all speakers. I remember reading a paper by a very respected researcher who was trying to get apes to do some number work through

African grey parrots are clear speakers, understand numbers and actually produce human words in context.

The grey's brain is densely packed with neurons and has seven areas for vocal learning.

traditional training. It took this poor ape 6,000 trials. But Koehler's birds succeeded almost immediately. Dietmar Todt was working with these same birds and teaching them little communication dialogues through observational learning. Things like "Hello, my name is Laura. What's your name?" These were the birds and techniques to use.

You point out that the grey parrot has a brain the size of a shelled walnut. So why are they smart?
Otto Kalischer, who was studying grey-parrot brains, found something he called the hyperstriatum, which seemed to be the center of intelligent behavior. His papers were in German. Very few Americans had read this stuff. I'd give a seminar, and colleagues would say there's no brain there, there's no cerebral cortex, how can greys do this? I would say, go find a cortex. In 2005, a paper came out with 21 authors—Erich Jarvis was the lead author—basically saying, "Hey, for birds there's this brain area that functions like the cerebral cortex. It doesn't look like a cerebral cortex at all, but it functions like ours."

And among birds, the parrot was special?
In the larger parrots, like the grey, the relative size of this chunk of brain is enormous compared to other birds, like pigeons. And there were seven vocal learning centers that seemed to have some analogy to the seven vocal learning centers in the human brain. In 2016, another group found that the neural density of the parrot brain is enormous and much denser in this

cortexlike area than in primates of comparable size. So now you have the reason they are smart: They have a brain area that works like a cortex, that's densely packed with neurons and its relative size is comparable to our cortex.

You named your famous grey parrot Alex, an acronym for Avian Language Experiment. When you decided to work with greys, you went to a store that sold parrots but let the clerk pick him out for you.

Right. I didn't want anybody to say I had chosen a special bird bred for intelligence. After that, he was in a lab with a small army of students. He had human companionship roughly eight to 10 hours a day. We treated him like a toddler. We talked to him. We described what we were doing. We constantly talked to him about things that were going on. If we were doing things like slicing green beans, we'd say "Ooh, look. Here's your green beans. They're beans. The color is green. Look."

You spoke with him like you were talking to a child.

Exactly. No subsequent bird has had that intense interaction, because it has had to share us with Alex and other birds. Griffin had it a little bit, because he had his own room for his first couple of years. He then spent most of his time around Alex, who dominated him completely. We tried to teach Griffin "What color?" and Alex would say, "No! Tell me what shape." And Griffin would look at me and look at Alex and shrug his little grey shoulders.

Irene Pepperberg shows colors and letters to the grey parrot, Alex, in her lab.

Parrot Aviary

Irene Pepperberg has spent almost four decades working with four highly trained greys to study parrot cognition and intelligence.

Alex (1976–2007)
possessed more than 100 vocal labels for different objects, actions and colors and could identify certain objects by their particular material. He could count object sets up to six and was working on seven and eight. Alex was learning to read the sounds of various letters and had a concept of phonemes, the sounds that make up words.

Arthur, aka "Wart" (1998–2013)
was the youngest bird at the lab. He loved to play with all of his toys and would say "want some" whenever the research assistants ate lunch. His favorite word was "tickle."

Griffin
was hatched in 1995 and began his training in the lab at a very early age. He has labels for many objects in the lab and is learning shapes and colors. Griffin loves grapes, bananas and apples. He loves to play with and shred his paper toys.

Athena
was hatched in 2013. She is still learning to communicate with humans and master the basic concepts of math.

So Alex would spontaneously speak entire sentences in context?
Yes. He asked questions. If we brought new things into the lab, he would ask, "What color?" "What shape?" "What's that?" or "What's here?" That's how he learned "carrot," "orange" and "gray." When he died, he was learning "brown" because he wanted to learn the color of the boxes he liked to chew.

Did You Know?
The large sulphur-crested cockatoo, found in wooded areas of Australia, New Guinea and South Africa, can learn about 20–30 human words. More commonly, they are known for babbling a stream of sound resembling speech.

You mentioned that he learned to recognize the color gray by looking at himself in a mirror?
Yep. The other birds just haven't had that experience, but they're also smart. We're doing things now with Griffin that 5-year-old kids can't do.

How old is Griffin?
He's 23. Alex was 30 when he died.

Are parrots smarter than children?
Little kids have a lot of growing up to do. There are precursors of these behaviors in very young children, but not this kind of logic. They are as smart as chimpanzees and dolphins. In a few studies, greys are better than apes. On some tasks, greys are as smart as 6- to 8-year-olds. On other tasks, they're at the level of maybe a 4-year-old child.

How do you make these comparisons?
We started by training the birds to communicate in English. Then we used English to examine their abilities the same way you would test small children. We trained Alex to identify colors, shapes and materials. Then we could show him an object and ask, "What color?" "What shape?" "What toy?" We could show him two things and ask what's the same and what's different. He understood concepts of similarity and difference at a very complicated level.

You also tested for number comprehension?
He could label sets of one through six. You could show him any set of junk, ask "How many?" and he'd tell you the number. You could give him a mix of red and blue balls and blocks and ask, "How many red blocks?" or "How many blue balls?" He'd tell you. We would give him trays of this mixed grouping of things and ask him, "What color six?" He'd see a tray with three of something, four of something, six of something all mixed up and he would have to find the set of six and tell us it was blue. He was right about 90% of the time. By the time he died, he could do simple addition. What blew me away was when he expressed the concept of none.

"He had human contact eight to 10 hours a day. We treated him like a toddler and talked about things that were going on." —Irene Pepperberg

You've said that he even invented his own words.
He said "banerry" when he was only a couple of years old, for apple. We were trying to teach him apple, but imagine trying to say the *p* sound without lips. Not easy. But he already knew "banana" and "cherry." Think about an apple—it looks like a big cherry and tastes a little like a banana.

A lot of scientists believed parrots are just mimics.
Colleagues thought I was crazy. My first grant application came back, literally asking me what I was smoking.

The talkative rose-breasted cockatoo, which thrives when stimulated with toys, can live as long as 80 years.

The macaw parrot can mimic human speech. Some, like this gorgeous blue-and-gold one, can learn complete phrases and reproduce sounds.

A pet macaw learns to talk most readily when food is used as a reward.

But when Alex died, there were tributes to him in scientific journals. What changed?
People finally accepted the work. Now a lot of scientists work in the area.

How are African parrots currently doing in the wild?
These birds are now among the most endangered species in the world. Huge flocks of greys used to live all across equatorial Africa. The numbers have gone down so dramatically that there are almost none in eastern Africa. Numbers in western Africa have gone down precipitously. People must learn not to buy wild-caught birds. We need to save their habitat. We need to protect them against poachers. When people realize how similar these birds are to us, they start to appreciate them more. So I'm hoping that people will use my work to do what's necessary to keep grey parrots from becoming extinct in the wild.

To that end, what does your research tell you about human and animal minds?
There is much less difference between animal and human minds than people once thought.

Obviously we differ in many ways, but sometimes we are the inferior species. Parrots see ultraviolet. We don't. Snakes can sense differences in temperature of 1/100th of a degree. We can't. Dolphins have sonar. We don't. There are differences. But the point is, there are many similarities—and we have to appreciate the fact that we are all denizens of this world.

Koshik learned to pronounce words in Korean by putting his trunk in his mouth.

From the Elephant's Mouth

Parrots are especially gifted, but a number of animal species have learned some human speech.

We may not live in a cartoon world like Bojack Horseman, but a number of animals have appeared to communicate with human speech. One of the most notable is an Asian elephant named Koshik (above), from a zoo in Seoul, South Korea. Koshik was isolated from other elephants and dealt solely with humans for seven years. He reportedly reproduced five words in Korean: *annyeong* for "hello," *anja* for "sit down," *aniya* for "no," *nuwo* for "lie down" and *joa* for "good." He "spoke" by placing his trunk in his mouth to modulate the shape of his vocal tract during pronunciation, a phenomenon never before observed, according to Angela Stoeger, an expert on mammal vocal reproduction at the University of Vienna. Stoeger notes that Koshik matched Korean pronunciations "in such detail that native speakers (could) readily understand and transcribe the imitations." And toward what end? Without other elephants to keep him company, one hypothesis holds, Koshik had come up with a creative and novel way to cement his interspecies social bonds.

Koshik is not alone. Two other mammals known to imitate human speech when isolated from compatriots were a white whale named NOC, a resident of the National Marine Mammal Foundation in San Diego, and an orphaned harbor seal named Hoover, who was picked up as a pup and raised by George and Alice Swallow in Cundys Harbor, Maine. Finally sent to the New England Aquarium in Boston, Hoover uttered "hey," "hello there" and "get outta here" with a notable Boston accent.

Elephants may paint specific images with some training.

Elephants may not have hands, but many can deftly paint with their trunks.

Animal *Artists*

Paintings and sculptures produced by creatures are featured in YouTube videos and sold in galleries and zoo shops—but are the creations real art?

When the male great bowerbird wants to lure a mate, he creates a fantastical architectural pathway adorned with brightly colored objects and stones. Female great bowerbirds decide on a mate only after touring the offerings and choosing what they deem the best.

In Baker City, Oregon, the annual Great Salt Lick Art Auction features salt licked by livestock into "tongue sculptures" that resemble works of modern art.

A rabbit has gotten into the act. The notorious internet star and meme, Bini the Bunny, who now lives in Los Angeles, routinely grabs a brush in his mouth and applies layers, dots and globs of paint to canvas walls.

And at the Institute for Marine Mammal Studies in Gulfport, Mississippi, a dolphin named Chance has shown a real passion for painting with color and a foam pool-noodle brush. "He likes to do different strokes. He likes to go side to side, up and down. And I think his favorites are a little swirl," his trainer, Lisa Crawford, said on the local South Mississippi news station WLOX.

Rabbits are doing it. Birds and cows engage. Across the species, many have the impulse to paint, sculpt or design.

AT THE ZOO

At zoos and sanctuaries worldwide, animals ranging from lions, lemurs and elephants to orangutans, penguins and chimps are painting with brushes and sticks or are walking, strolling or running across canvases primed with nontoxic paint. The resulting artworks are often sold by these institutions to benefit the animals they serve.

At the Houston Zoo, a YouTube video shows a parrot dabbing at a canvas with a brush held in its beak, a chimp applying paint with a stick held in its paw, and elephants smearing paint on canvases with brushes clutched in their trunks.

At Florida's Palm Beach Zoo & Conservation Society, where some 24 animals from small minks and armadillos to jaguars and bears are encouraged to do artwork, painting offers the animals access to interesting and complex activities that help break up their daily routine, according to Amy Anderson, associate curator of herps and program animals.

"Painting is both training and enrichment for the animals," says Kevin Hodge, the Houston Zoo's general curator. "Some animals enjoy the smell and texture of the paint, as well as the visual enrichment of seeing the colors on the canvas. We've had some animals that loved rolling in the paint. And for some animals, like our cougars, participating in art gets them a food reward."

Training an animal to paint can take a bit of patience. At the Palm Beach facility, Anderson and staff teach animals to paint through the use of a target that "looks like a giant Q-tip." Otters and jaguars, for instance, learn to touch the target with their mouth or their hip. Should they get close to the target, the animals are reinforced with food or whistles "to let them know

If you want a painting created by a jaguar, you can custom-order your canvas at the Palm Beach Zoo.

they've done a good job," Anderson explains. "When animals hold their mouths on the target for two seconds, you reinforce their behavior again."

In other words, you get them to learn a new, unique action—holding a brush or touching a canvas. "And [eventually] the time they hold it gets longer," she says. Before you know it, the animal is involved in producing a piece of art.

It can be tricky to know when an animal artwork is truly complete. "We keep a close eye on how interested the

Not just an artist, Bini the Bunny can also slam dunk a basketball through a mini hoop.

Denver, a blue-and-gold macaw, is a popular artist at the Houston Zoo.

animal is," says Hodge."When they're done, we're done. A painting may take a couple of sessions to complete. It all depends on the animal, but it's usually not more than once every few weeks." Anderson says the Palm Beach Zoo's animals may paint "once every three months, for a few minutes."

To an interested public, some animal artists are more in demand than others. "A lot of guests love otters, and art from some of the larger carnivores—jaguars, tigers, panthers, bears," according to Anderson. "Probably next are some of the primates." Those interested in the artworks can custom-order a piece of art from the zoo's website—choosing a mammal, reptile, or bird artist, as well as specifying canvas size, background color and three paint colors.

ARE THE ANIMALS ARTISTS?

But are the animals really creating works of art? Anderson certainly thinks so. As an example, she cites the zoo's otters. "They're wonderful artists—you squirt the paint on a canvas on the ground, call in the otters, and they'll walk on the paint. They're superactive animals so they tend to go over the canvas again and again, and that creates nice paw prints."

But more importantly, she says, these animals are earning their keep. "One of the biggest things for the public to know is that the money from these sales is going right back into the care of our animals and giving them excellent welfare. The animals are, in effect, fundraising for themselves."

Consumers can order animal art from a lion, orangutan or elephant at the Houston Zoo's website, which proclaims: "From paw prints to broad brushstrokes, each artist's creation is one of a kind." For $250, the zoo includes a flat, unmatted 6-by-20-inch canvas; a photo of the animal artist; and a short biography. At Florida's Palm Beach Zoo & Conservation Society, animal art ranges from $32 for a 5-by-7-inch print to $106 for a 16-by-20-inch canvas.

As to copyright issues, animal art seems to be fair game for anyone who wishes to reproduce it. The law was established in 2014, after a Celebes crested macaque in Tangkoko Batuangus Nature Reserve in Indonesia took a series of selfies with a photographer's equipment. In August of that year, the U.S. Copyright Office declared that those selfies and any other animal-made art fall outside the province of copyright law, which applies only to humans. If elephants, chimps, dolphins and other highly intelligent species are ever given human status, look for the situation to change.

This abstract painting was created by Congo the chimpanzee (seen below) in the London Zoo during the late 1950s.

'Cézanne of the Ape World' Was an Art Pioneer

Congo, a chimpanzee from the London Zoo, was the first nonhuman artist to make a critical splash.

It was in the late 1950s that Desmond Morris, the famed zoologist and anthropologist, hosted a TV show called *Zoo Time*. The hottest star on the lineup might have been Congo, a resident since 1954 of the London Zoo. Morris said that Congo at first just splashed paint on paper in a haphazard way. But slowly, that changed. "I was amazed. He focused on what he was doing. Every line he made logically followed the last one," he told the *Guardian*.

It was Morris who first reportedly gave Congo a pencil and observed the chimp's ability to draw circles and balance color and structure on the page, which evolved over time. Eventually, Congo produced a series of radiating-fan drawings (see above).

Morris observed that Congo always knew when a drawing was done, and once it was he would paint no more until a blank sheet of paper arrived. If an unfinished painting was taken away, Morris added, Congo would "throw fits."

The prolific chimpanzee produced more than 400 paintings by the time of his death from tuberculosis in 1964 at age 10. Even Pablo Picasso is said to have owned a Congo painting, which was rumored to hang on a wall of his Paris studio for years. Picasso wasn't alone. On June 20, 2005, Congo's tempera paintings were first put up for auction alongside works by Andy Warhol and Pierre-Auguste Renoir at Bonhams, where three of them sold for more than $25,620.

CHAPTER 4

Lessons From the Wild

From life after a nuclear disaster to elaborate underwater songs, there's much we can learn from our fellow species.

Canines formed pack families in the ruins of the nuclear accident.

Stray puppies play in an abandoned, partly finished cooling tower inside Chernobyl's exclusion zone.

Dogs of Chernobyl

After the meltdown of the nuclear reactor in 1986, people evacuated, but pups stayed behind. Here's what happened next.

Andriivka, Chernobyl, Kopachi, Poliske, Pripyat, Tarasy, Velyki Klishchi, Vilcha, Yaniv. These are modern ghost towns created by the same nuclear disaster. Once lively northern Ukrainian municipalities that housed farmers, artisans and employees of the Chernobyl nuclear power plant, they were deserted following a catastrophic accident that happened at the nuclear plant on April 26, 1986. The trouble started when a power surge occurred during a routine safety test, sparking two enormous explosions in the plant's No. 4 nuclear reactor. The explosions were strong enough to blow the reactor's 1,000-pound roof off, allowing massive amounts of radiation—the equivalent of 400 Hiroshima bombs—to leak into the air, water and soil, and into plants, animals and people, as well.

While it was clear inside the smoldering Chernobyl nuclear power station that an accident had occurred, the rest of the world was kept in the dark. Ukrainian officials evacuated Pripyat within 36 hours, but it was a full two days later that workers at a nuclear power station nearly 700 miles away in Sweden rang the international alarm bell. They'd detected radiation on their clothing that didn't come from their power plant; it had come from Chernobyl. Dire news of the disaster prompted officials to draw a 30-kilometer radius around the plant referred to as the exclusion zone. People living inside had to drop their belongings and leave their contaminated homes forever.

The fate of the animals inside that contaminated area is far more complicated. As the people cleared out, government officials called "liquidators" were ordered to shoot and kill evacuees' pets (and many wild animals) in a bid to prevent the spread of radiation. They buried the bodies in enormous heaps beneath contaminated soil.

The dramatization of the killing of the dogs in the 2019 HBO series *Chernobyl* shocked viewers. Craig Mazin, the series creator, tweeted, "I know that was hard [to watch]. Just so there's no confusion—the story of the liquidators is real. It

"Man's plight makes you sad, but the plight of the animals is even worse."

—Svetlana Alexievich, *Chernobyl Prayer*

This stray dog stands at the monument by the new, giant enclosure that covers the devastated reactor at Chernobyl.

happened. And we actually toned it down...."

But the liquidators didn't completely wipe out Chernobyl's dogs. Maybe as few as a handful of survivors forged tedious lives on the exclusion zone's emptied streets. Many didn't survive puppyhood, due to disease and starvation, but others formed packs and scavenged for resources such as food and water.

At the same time, thousands of soldiers were recruited to reduce some radiation in the zone by removing and burying contaminated materials, like soil. They also worked to ensure Ukrainians obeyed the restrictions on entering the contaminated area.

Despite this, a few people refused to leave the zone in 1986, and through the 1990s, several hundred more returned to continue lives left behind. The returnees were called "samosely" or "self-settlers."

A LIFE ON THE EDGE

By 1999, an estimated 612 samosely were living in Chernobyl with perhaps the same number of dogs. It's not clear how many dogs the liquidators killed, but samosely are thought to have cared for at least some that evaded the post-disaster killings. For the dogs, who were barely hanging on, forming cautious alliances with these people meant getting leftovers and a better shot at survival.

After a few years, the original returnees were joined by thousands of workers charged with removing contaminated materials, mostly soil, and constructing a new concrete building to encase the dangerous old reactor and prevent more radiation from leaking out.

This aerial shot shows Chernobyl's nuclear reactors with canals in the spring.

By 2010, the exclusion zone saw an influx of even more workers recruited for cleanup and the construction of a new confinement structure. These power station employees have formed a special bond with the dogs, says Jennifer Betz, a volunteer veterinarian tending to Chernobyl's dogs. It's common to see Chernobyl workers splitting their lunch sandwich with a favorite feral hound.

Despite the power plant workers' hospitality, the dogs still lived life on the edge. Disease and hunger persisted. And while hundreds of new pups were born each year, just half of them survived.

In recent times, tens of thousands of tourists have come to the area, drawn in by Chernobyl's radioactive legacy and the lure of disaster porn. Thanks in part to the popularity of the HBO series, Kiev's tourism and promotion department expects the number of visitors to Chernobyl to continue to grow.

Ukrainian officials, ready to capitalize on a return of humans to the formerly abandoned zone, were, naturally, concerned about rabies among the 1,000 unvaccinated dogs. Though the evacuated area covers about 2,600 square kilometers, dogs are not widely dispersed. Instead, they concentrate where humans live: A few hang around samosely dwellings, more than 100 can be found at the Chernobyl nuclear plant site and many more have spread throughout the city of Chernobyl, where most of the zone's population is concentrated right now.

SAVE THE DOGS

In 2017, more than three decades after the meltdown, volunteers began to arrive to help the approximately 1,000 dogs living in the zone. "They've been breeding and breeding for 30 years," says veterinarian Betz. Coming into the zone with a crew of volunteer vets, vet techs and people affiliated with an American nonprofit group called Clean Futures Fund, she began providing canine medical care. This includes first aid, spaying or neutering, vaccination and deworming to protect against parasites.

"Our main goal is to provide a better life for the dogs," says Betz. "Many die from disease and starvation in winter. We couldn't bring them out of the zone, so we came here to treat them."

It remains challenging work. Getting a hold of the dogs isn't easy; you can't trap them like other street dogs. "They're too smart," Betz says. The friendliest young puppies can often be scooped up by hand, but adult dogs must often be blow-darted with a sedative cocktail. After that, sterilization surgery commences,

Animal advocates cuddle with puppies at a makeshift vet clinic operated by the Dogs of Chernobyl initiative.

vaccines and deworming medications are administered, a radiation-reading dosimeter is clipped onto the ear and the dog is released back on the streets exactly where it was found. Like all feral dogs, the Chernobyl canines don't mix well with dogs from other packs, so it's important the release location is precise to prevent skirmishes.

A very small number of Chernobyl puppies have been adopted out to individuals within the Ukraine and in North America. Yurislav, aka Yuri, is one such dog. "He is very well tempered, house trained, knows several tricks and commands, and is great with children," says Tim Mollohan, Yuri's adopter. Mollohan is an American nuclear professional from Georgia who oversaw the containment of spent fuel at Chernobyl's nuclear power station from 2017 to 2018. There, Mollohan says, he became a friend to many of Chernobyl's street dogs. And he was working in the highly contaminated region when he got the chance to adopt one of these dogs and bring him back to the U.S. after his work at the power station was complete. He says he wasn't worried about radiation. "I am familiar with radioactive contamination and how it passes through the body. I trusted the Clean Futures Fund team and their decontamination abilities," he says. "I knew my family would be safe, and little Yuri had not been exposed long enough to significant levels of radiation."

According to Lucas Hixson, Clean Futures Fund co-founder, efforts by his group and others have led to healthier and longer lives for the dogs. Controlling for potential rabies outbreaks has also led to an improvement in public health. "Ukraine gets its rabies vaccines for humans from Russia, but because of the war they haven't received an adequate supply in years," says Hixson. "This means a person potentially exposed to rabies will have to travel at least three to four hours to find a vaccine."

Today, thanks to sterilization, fewer dogs are born, and those that remain have easier lives than the pets who escaped the liquidators decades before.

A BOON TO SCIENCE

While it's clear the volunteers' project is increasing the Chernobyl dogs' well-being, it has come with an unexpected scientific benefit. Scientists are able to study the dogs while treating them, drawing blood, saliva, hair and fecal samples and bringing them back to the lab. Researchers are using the opportunity to answer two fundamental questions about the Chernobyl dogs, says Tim Mousseau, an evolutionary biologist at the University of South Carolina: "Who are they, and what can they tell us?"

Some of these answers already seem clear. The dogs, who live just two to four years due to a hard life on the street, share a similar look that is different than the more diverse array of pet dogs found in surrounding villages: They are often robust and shepherd-like with large and erect ears, ranging between 45 and 65 pounds, with either short or long hair. And though they've been getting generous handouts from the samosely, workers and tourists, most of the dogs are rather feral, scurrying away when humans approach.

Their survival strategy has left them "rather plump, as far as street dogs go," Betz says. She and the other volunteers have not yet observed evidence of physical mutations linked to radiation like tumors, cataracts and smaller brains, all seen in other animals, such as birds, in heavily radiated areas like the nearby Red Forest.

Mousseau says DNA analyses of hundreds of their blood samples are underway, done in collaboration with geneticist Elaine Ostrander at the National Institutes of Health. Mousseau and Ostrander are now trying to prove that the dogs roaming around the long defunct, but still dangerous, Chernobyl nuclear power station are descendants of the disaster's original canine survivors.

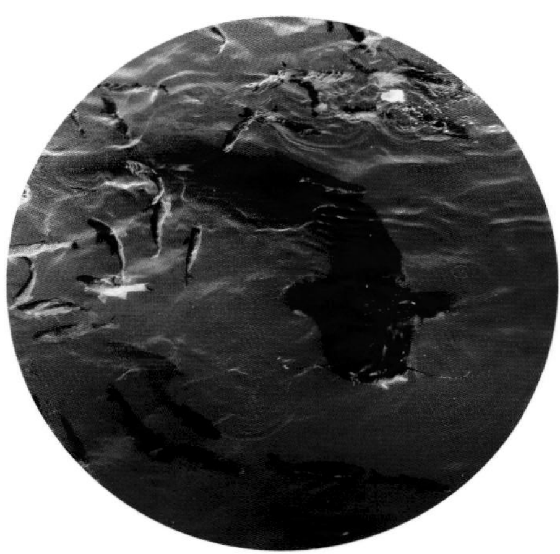

Large catfish live in the cooling ponds at Chernobyl, but scientists say their surprising size is not the result of mutation.

Did You Know?
More than 30 years after the devastation at Chernobyl, the red fox maintains a home in the area. It is one of 14 mammalian species observed often enough by scientists to be considered part of a robust, reproducing population.

If scientists can trace the dogs' lineage back to this small founder group, it would open up the first-ever opportunity for scientists to study the long-term health effects of initial exposure to high-level, and now low-level, radiation on a genetically uniform mammalian population, Mousseau states.

"This would mean we can examine questions of adaptation and evolution that wouldn't be possible if there was gene-flow coming in from other areas," says Mousseau. "If these dogs are descendants, we're looking at a small and isolated population, and this gives us more insight into their evolution and adaptation to a stressful environment." The generational genetic-level effects of radiation exposure are much more challenging to study in outbred populations, like humans, he adds.

A BETTER FUTURE

Most of the radiation from the Chernobyl blast settled over more than 160,000 square kilometers across Belarus, the Russian Federation and Ukraine, with lower levels of fallout across Europe. Today, radiation levels throughout Chernobyl and its surrounding areas vary. Some regions, such as the Red Forest, still clock in very high dose rates. "This contamination is found throughout the environment, so animals are still exposed to radiation through the soil, water and air, as well as the plants and animals they eat," says James Beasley, an ecologist at the University of Georgia who studies wildlife in the exclusion zone. "Because of this, many animals continue to accumulate very high levels of contamination within their body tissues," he says.

So far, most dogs' dosimeters reveal low levels of radiation exposure, Betz says. However, she and Mousseau plan another collection of readings to look for any patterns as to what a "normal" radiation dose is for a Chernobyl dog. If scientists find that the genomes of the Chernobyl dogs are unusual, it could determine if genes mutated by radiation exposure can be passed down—or build up—over generations.

—Heidi Hutner and Erica Cirino

Heidi Hutner, PhD, is a writer, documentary filmmaker and professor at Stony Brook University in New York whose work covers nuclear history, ecofeminism and environmental justice. Erica Cirino is a writer, artist and wildlife rehabilitator who covers stories about the environment.

Wild horses live and thrive in the fields irradiated by the meltdown at Chernobyl.

The horses at Chernobyl still face many threats in the form of poachers.

How Other Animals Adapted to Post-Disaster Life

Mice, bears, birds and more animals also found ways to thrive amid the destruction.

Dogs weren't the only species to survive the worst nuclear meltdown of all time. Among other survivors, you can find bears, bison, deer, moose and many bird species. To document the comeback, a study from 2015 revealed abundant wildlife, including gray wolves, in the 1,000-square-mile ecological zone. More recently, in 2019, researchers from Georgia planted fish carcasses as bait and then set up cameras to see which species could be observed. According to James Beasley, associate professor at the Savannah River Ecology Laboratory and the Warnell School of Forestry and Natural Resources and one of the researchers, the team saw 10 mammalian species and five species of birds. The results, published in the journal *Food Webs*, found that nutrients from bodies of water, especially rivers, can flow into damaged land regions, feeding species there on the ground. The Georgia researchers found that scavengers consumed 98% of the fish carcasses in a week. Species identified in the study include three types of mouse, the least weasel, American mink, Eurasian otter, pine marten, red fox, Eurasian jay, common magpie, raven, tawny owl and the white-tailed eagle.

There are more than 300 octopus species, including the blue-ringed octopus (*Hapalochlaena lunulata*).

Aliens of the Deep

The genius of the octopus comes from a brain and nervous system evolved to coordinate eight arms.

If you've seen an octopus horror film, you know this creature has many arms—eight, to be exact. But have you heard about its dazzling skin? In seconds, thousands of tiny hued pockets open and close, allowing the octopus to change colors for camouflage and communication. Skin texture changes too, making an octopus completely invisible from just a few feet away. One species, the mimic octopus, can impersonate more than 15 animals, including flounder and sea snakes. And if necessary, an octopus can shrink and squeeze through a hole not much bigger than its eye.

The evolution of humans and octopuses separated early on. Even so, these shimmering shape-shifters share some startling similarities with us. We are both soft-skinned predators with unique skills for competing in a dangerous world. Octopuses build dens. They use tools and spatial navigation. Some species of octopus play. And at least one can distinguish between individual humans. In recent years, scientists have even discovered octopus genes for sociability.

MULTITASKING HEROES

The fixed, catlike gaze of the octopus has spoken to fishermen through centuries. Yet the classic question about consciousness—what does it feel like to be an octopus?—is unanswerable. As philosopher and diver Peter Godfrey-Smith puts it in *Other Minds: The Octopus, the Sea and the Deep Origins of Consciousness*, if we can make contact "it is not because of a shared history...but because evolution built minds twice over."

Millions of years ago, the snaillike ancestor of the octopus lost its shell. That made it more versatile, but soft and vulnerable. To flee predators, it became a master of camouflage and escape, over time acquiring considerable intelligence. A human has about 100 billion neurons. An octopus has half a billion, a tiny percentage of ours, but about as many as a dog.

One huge difference is that most of its neurons are in its eight arms; with their own control centers, they

The giant Pacific octopus can tell the difference between a caretaker that harasses it and one that brings food.

Each arm of an octopus can sense and think independently of the others.

Octopuses depend on oxygen for visual information—and climate change may hamper their ability to see.

can touch, taste and even see and think independently, putting human multitasking to shame. An octopus arm can figure out how to open a clamshell while its brain scans for predators.

SQUIRTING PROTEST

Captive octopuses understand their situation, frequently trying to escape when they aren't being watched. They deliberately plug valves in their tanks to flood the lab and famously squirt at people who frustrate them. Gül Dölen, a neuroscientist at Johns Hopkins Medicine, observed, "The biggest difference between an octopus and a mouse is that octopuses are the dominant predator in their niche, while mice are a bit like the potato chips of the animal kingdom (eaten by almost everybody)." Octopuses are so bold, she notes, that they will "squirt you with water if you catch them trying to leave their tank and you nudge them back in."

Retired researcher Jean Boal has reported that when she fed octopuses thawed squid, a substandard meal, along a row of tanks, one octopus gave her trouble: It held the squid out in its arm, and watching Boal, slowly crossed the tank and shoved the squid down the drain (sound like a toddler?).

The giant Pacific, one of the more social species, can distinguish between people who cross it, as the late biologist Roland Anderson demonstrated in a 2010 study at the Seattle Aquarium.

Anderson's team had one experimenter harass eight giant Pacific octopuses with a bristly stick, followed by a good guy wearing the same blue uniform who brought food. Two weeks later, the octopuses would shrink into the corner, turn their suckers out to fight and blow jets of water at the bad guys—but greet the foodgiver by raising their arms.

It seems obvious the octopuses like the food man and dislike the stick man, but we can't know what the octopus feels. Emotions arise from social ties, and octopus society is limited.

Their existence is the opposite of our civilized lives. Most species of octopus are radical loners. They survive for only a year or up to four, depending on the species, without learning much from peers or bonding with a mate. Octopuses don't even bond to parents. Their mothers die after their eggs hatch and their fathers soon after that.

In a standard pattern, an octopus will make a den, live there for a few weeks, and then leave it to set up another. Males in some species compete for mates. A male might guard a female against other males, trying to keep her for himself for a week or so. But mating is often from a distance, handing over sperm with an outstretched arm. Males seem to learn their social rank, varying their displays depending on how likely they are to win a particular fight. They also have a variety of behaviors to avoid fights. "Whether they feel that rejection of a lost fight or a missing mate cognitively, I don't know," observed Christine Huffard, a researcher at California's Monterey Bay Aquarium Research Institute. Females mate with just about any male that comes by, even if another male is already guarding her.

But a few years ago, observers confirmed earlier accounts of oddly social behavior among a rare species, the larger Pacific striped octopus, which has been seen in groups of up to 40 off the Pacific coasts of Nicaragua and Panama. Mating pairs shared meals beak-to-beak and also grasped each other's arms and mated beak-to-beak—let's not call it kissing.

Current research is based on observations of just a few of the more than 300 species around the world, and new findings arrive all the time.

Did You Know?
The multiarmed squid, above, is a cephalopod like the octopus. Both animals change color for camouflage by controlling the size of their shells and both squirt ink, but the squid lives in the open ocean while octopuses inhabit ocean dens.

GIVING AN OCTOPUS ECSTASY

In recent years, scientists have gotten a bit of a surprise. They decoded the genome of the California two-spot octopus (*Octopus bimaculoides*), a big loner, and then discovered areas of genetic code, identical in the human genome, involving brain chemicals tied to social life.

The finding inspired Dölen's Johns Hopkins team to treat four of these octopuses to a dose of ecstasy (MDMA)—affecting those brain chemicals—to see if they warmed up. Humans react to the drug by touching each other more. So did the octopuses, who began to hug the cage and put their mouth parts on it, apparently to contact the octopus in the neighboring cage.

But as much as we love to see human traits in octopuses, their genius also lies in the ways they are not like us, Huffard observed. Octopuses have three hearts—two for blood flow through the main body and one for blood flow through each set of gills. Some incredible octopus species are able to "swim" through sand and come up a few feet away from where they went under.

Their intelligent arms can move even when severed. Some species can eject an arm, which slithers off and distracts predators so the octopus can escape. "They take in so much information and have such complex bodies to control, it's a mystery how they're able to process it all," says Huffard.

In *Octopus! The Most Mysterious Creature in the Sea*, Katherine Harmon Courage describes her experience eating live octopus selected from a tank in a restaurant in Flushing, New York. The cut-up arms writhe, suckers gripping the plate. Contemplate that each sucker contains neurons giving it the power of taste, she notes. As you eat, the octopus arm is tasting you.

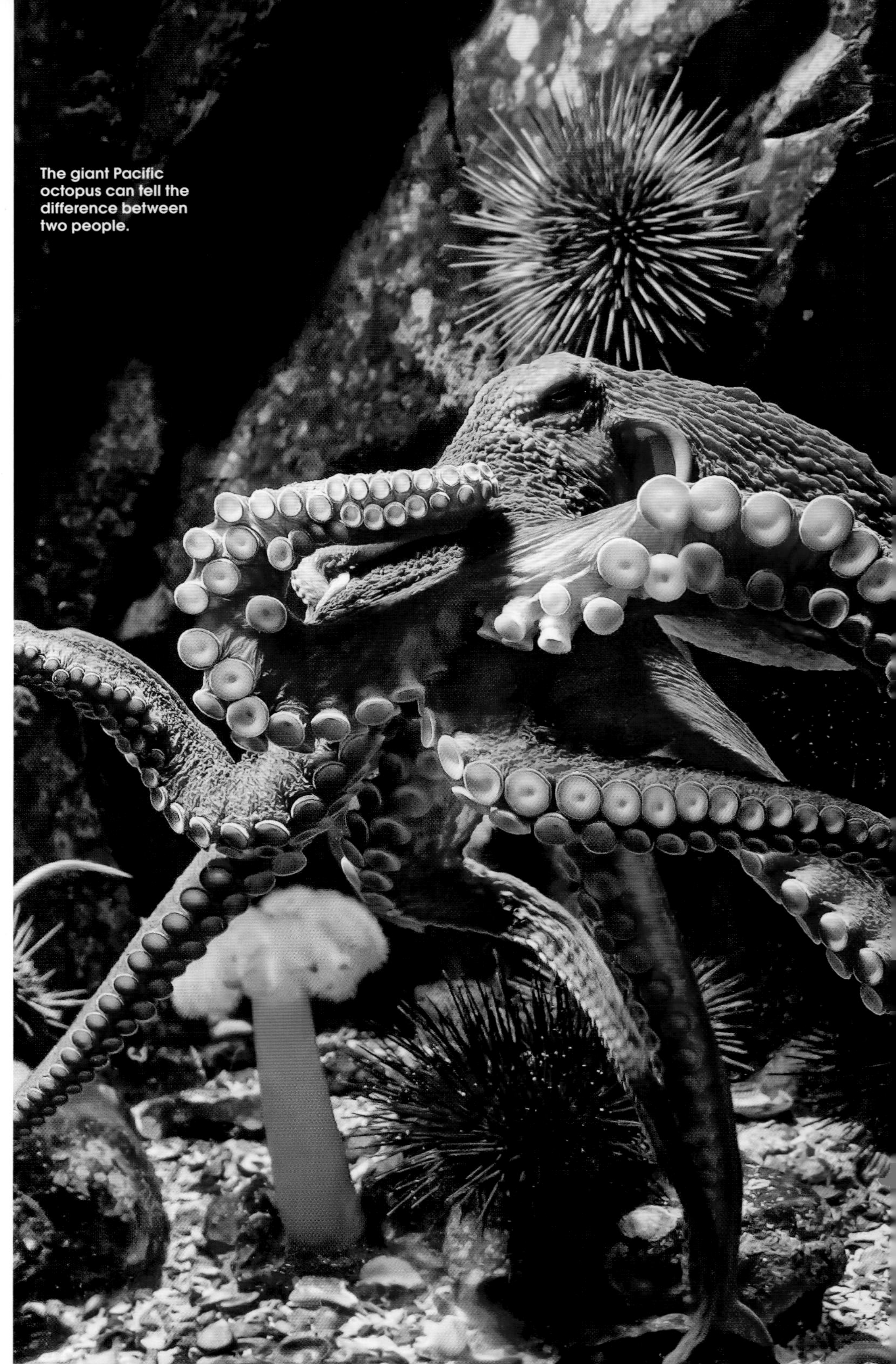

The giant Pacific octopus can tell the difference between two people.

Talk Like an Octopus

Octopuses communicate when fighting or mating. The exchange comes through movement and changes in the color of their skin.

The *Abdopus aculeatus,* a small intertidal octopus common throughout Indonesia, is a mottled gray, brown or ochre most of the time, blending into its habitat. In bright light, this species might glimmer with a green iridescence; near seagrass, it might turn brown with pale horizontal stripes.

But if a male is near a female with whom he wishes to mate, he's not trying to hide. Instead, he'll go pale and develop dark longitudinal stripes—high contrast for maximum visibility. It's the same pattern he takes on if he's about to fight another male. The female might also develop the striped pattern, though not as often. Before mating, the two octopuses sit upright, raising and lowering their mantles every few seconds.

Strangely, an octopus can show a display on just one side of its body, and vary its intensity. A related species, the cuttlefish, can send signals to entice a possible mate from one side and warn off competitors with a separate display on the other.

Although octopuses have camera-type eyes (with a lens, iris and retina) like ours, they can see only one color. Most octopuses likely see the world in shades of a blue-green, researcher Christine Huffard observed; that's why contrast is important to communicate. "Their skin, with its lightning-fast color- and shape-changing reflexes, reflects evolutionary adaptations over thousands of generations of...escaping from color-seeing predators. It's truly amazing to see it in action," she says.

The *Octopus tetricus,* which lives in the shallows of Jervis Bay, Australia, meanwhile, also adopts a "stand tall" posture, with a raised mantle when it's trying to show off.

Can humans and octopuses communicate? The naturalist Sy Montgomery famously communicated with an octopus named Athena at the New England Aquarium. But Huffard has reservations about trying herself. "I don't have the ability to make my skin give the signals they understand," she says. Tapping on the tank or flashing a light would stress them, and in the wild, chasing an octopus causes it so much stress it can kill the animal if it goes on too long. Handling can damage its delicate skin, and the handler risks a poisonous bite. (Like spiders, all octopuses have poisons in their venom to immobilize or help kill their prey.)

An octopus might reach for a diver's bare hand and pull the fingers. Is it stroking you affectionately? No, it's exploring, and likely to bite. In a blog to warn divers, Huffard wrote: "It has every reason to bite and few reasons not to. If you handle it, it may bite you out of defense because you are so much larger. If you try to lure it out of its den, it might mistake your hand for a meal."

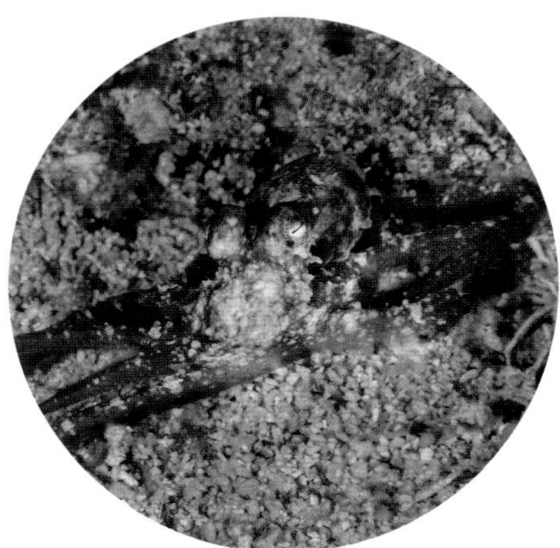

Algae octopus camouflage to appear like an algae-covered shell.

Bonobos can understand language, communicate with humans, play musical instruments and use tools.

Do Animals Have Beliefs?

Debates rage, but the latest research shows that species from birds and dogs to chimpanzees practice complex, flexible strategies based on motivations and expectations.

Every day shortly before I came home from work, my wife recalls, my dog Connor would leave his position on the couch beside her and move to the armchair near the front door. He'd lie there on the seat, stretch his paws across the arms and stare intently at the door. Based on our routines, he was convinced I'd walk through that door any minute. When I arrived, his tail started wagging frantically, his entire body started shaking, his neck craned forward and his eyes went wide with excitement, based on his anticipation that I would sit beside him pronto and start petting him. When I did sit down, he invariably flipped on his back after a few moments, expecting a belly rub. Minutes later, when I headed to the closet, he knew that I was getting his leash, and that we were heading off for his daily great adventure, our long evening walk. By the time I grabbed the leash, he had already leaped off the chair and run to the front door, where he stood with his nose pasted against it.

In those few moments, my dog displayed a series of strong beliefs, and he built the most important part of his day around them. That's my opinion, anyway. In recent years, the question of whether animals really have beliefs has sparked some feverish debate. But two German cognitive scientists/philosophers, Tobias Starzak, a graduate student, and Albert Newen, PhD, at the Institute of Philosophy II at Ruhr-Universität Bochum, have published a study that scrupulously tears down the naysayers' arguments, which hold that without a true language the concept of belief becomes meaningless or at least vague for species aside from our own. But, the researchers say, just because we don't know what's in animals' minds and don't share their belief system, doesn't mean they don't have beliefs.

Some animal species have a theory of mind—they understand that others have different thoughts, beliefs and desires.

Their argument in a nutshell: While we can't see inside an animal's mind, we can observe that animal's actions. "And then you see if ascribing beliefs to that behavior provides a good explanation for the behavior," Starzak explains.

To have a real belief, an animal must start with real information about the world and then act on it with intent and in a flexible way. For example, one dog might think something like, "I just saw a squirrel run into that tree, and I believe he is still in that tree, so I'll stand under the tree and bark, and maybe he'll come down." There's a reason he's doing what he's doing, based on both the belief that the squirrel is up there and his motivation to get the squirrel.

Another dog, which had been attacked by a rabid squirrel, might be motivated very differently and run away from the tree, based on the same belief that the squirrel is in the tree, as well as a different belief that it might hurt him. In other words, real belief can lead to different actions depending on an animal's motivation; this decoupling leads to flexibility of thought and action, Starzak explains.

Here are some of the most prominent and fascinating examples of animals putting such flexibility into action based on definitive beliefs.

AVIAN STRATEGISTS

We may think of them as simple tweety-pies, but different bird species have a dazzling array of strategies for foraging and self-protection that appear to involve definitive belief systems.

• **Piping Plover** This small, sand-colored bird has mastered an impressive ploy that scientists call the "broken wing display," which certainly *seems* to be the product of a sophisticated belief system. When a creature approaches a plover's nest on foot, if she deems it a potential predator, she lands on the ground and feigns having a broken wing. She hops pathetically away from the nest as if she can't fly, presenting herself as an easy morsel to capture and devour. When she has drawn the predator far from the nest, she simply flies away.

This strategy appears to involve several beliefs—that the particular approaching creature is a predator, that the strategy will save its offspring and

Piping plovers camouflage their eggs in depressions that are lined with pebbles and shell fragments.

Piping plover couples are monogamous and share child-rearing duties.

On some cognitive tasks, rats can do as well as a human 4-year-old.

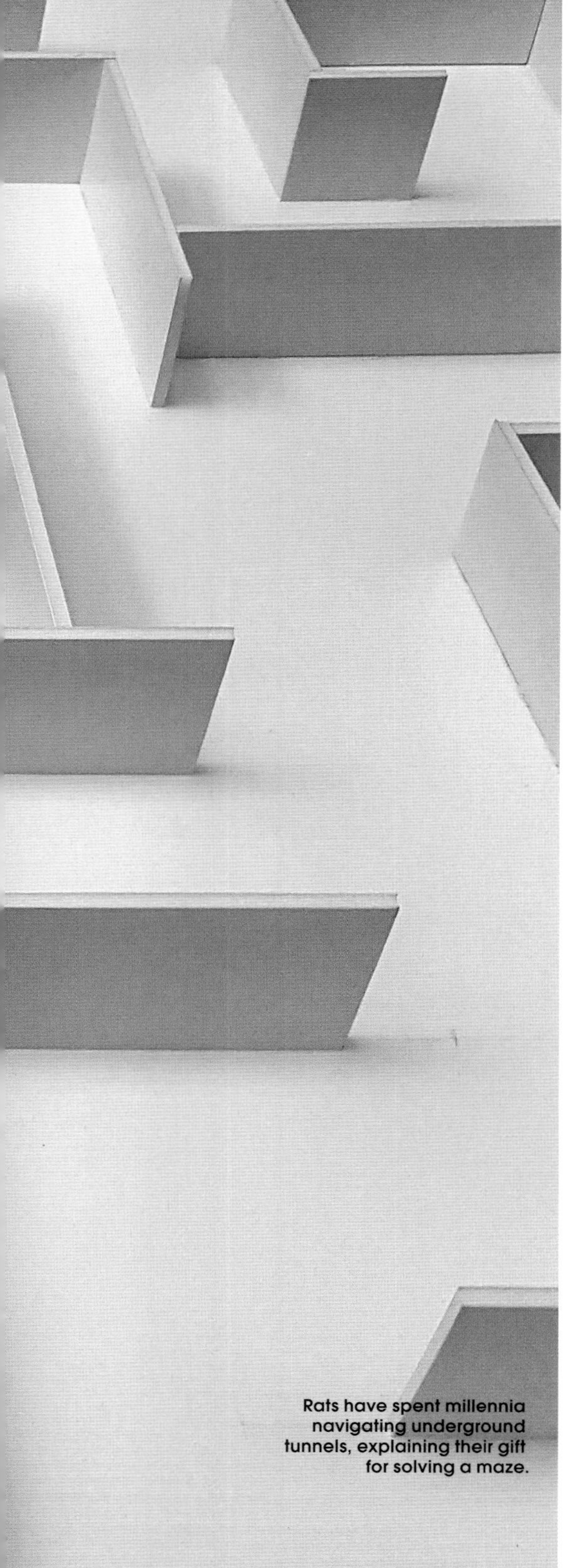

Rats have spent millennia navigating underground tunnels, explaining their gift for solving a maze.

that it will accomplish this by creating a false belief in the presumed enemy. "It sounds amazing, not only that the piping plover has a belief system, but that she wants to manipulate the behavior of the predator by making it believe something false. So that's super complicated," says Starzak. "But it's too automatic—it's the only strategy they use in such situations, so I think it's hardwired into them. It doesn't demonstrate the flexibility that would tell you the birds have a belief system."

• **Corvids** This family of clever crow varieties, including ravens, rooks, jackdaws, jays and magpies, has many sophisticated strategies for varying situations that demonstrate beliefs. In one experiment, ravens confronted a tube containing floating goodies just out of reach. They figured out that dropping nearby rocks in the tube would bring the water level up, delivering the treat.

In addition to getting food intelligently, corvids protect it intelligently. If one senses another corvid (or a different predator) nearby when hiding food, it comes back ASAP and re-hides the food. And here's the best part, according to Starzak: Their belief that their food is at risk comes not from their experience of being stolen from, but from their experience stealing from other corvids. "It's like, 'I'm not going to let them do to me what I did to them,'" observes Starzak.

• **Scrub Jays** These nonmigratory, urban jays were the stars of landmark research by cognitive psychologist Nicola Clayton. Clayton observed the intricate way they dealt with food they amassed. If they were hungry, they ate it. When they weren't, they systematically hid the leftovers—worms in one place and peanuts in another. If they became hungry in the following hours, they first went to where they hid the worms, which they preferred, but if more time passed and they presumed the worms would no longer be edible, they'd hit up the peanuts. "What best explains these changes in behavior, this flexibility, is the birds' beliefs about the worms being spoiled and about the location of other food items," says Starzak.

• **Parrots** Animal psychologist Irene Pepperberg's famous 30-year study with Alex the grey parrot showed he had astonishing flexibility of thought. She trained him to become the first animal to understand and communicate

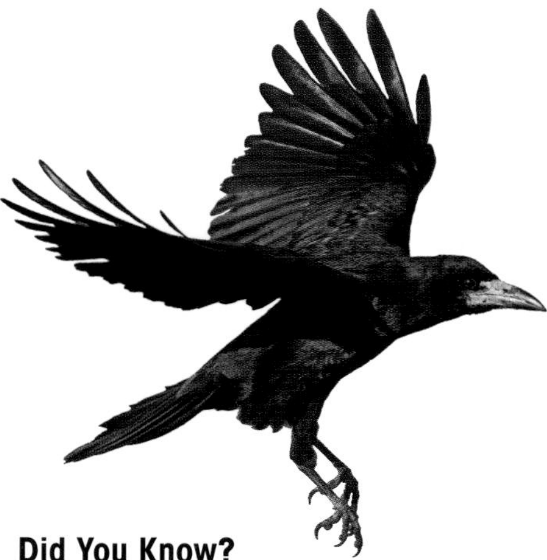

Did You Know?

The rook is the most abundant bird in the crow family. These intelligent corvids have been observed dropping walnuts onto busy roads and using traffic to crack them open. They can use and even modify tools and have the ability to plan ahead.

concepts, grouping objects into categories and then differentiating them by properties such as size, color and temperature. Alex learned to represent an object as red in contrast to four other colors; as round in contrast to three other shapes; and wooden in contrast to two other materials. "He could communicate, say, that one object was round, yellow and made of wood, while another was yellow but not round and made of something else," says Starzak. (For more, see page 160.) Pepperberg claimed that Alex could identify 50 different objects, recognize quantities up to six, distinguish seven colors and five shapes, and understand the concepts of bigger, smaller, same and different. Alex passed increasingly difficult tests showing he had incorporated the concept of object permanence into his belief system.

THINKING RATS

In a fascinating study reflecting tremendous associative flexibility, rats were trained not only to orient themselves in an eight-armed maze, but also to recognize where different foods would be found at three different times of day. They learned to understand that if they found normal food in arm three in the morning, they could find chocolate in arm seven at midday. "The rats were able to make complex cause-and-effect judgments of time, location and motivation chains of events," says Starzak. "We could justify ascribing to them this specific though complex belief: 'If I find normal food in arm three at 6 a.m., I will find chocolate in arm eight at midday.' Studies have shown that many rats can orient themselves spatially in labyrinths based on landmarks, are sensitive to findings at each time point and can update their expectation according to these findings, integrating all of this what-when-why information. They are very smart animals."

When environments change and foil their expectations, animals learn to change their habits and behaviors.

FLEXIBLE PRIMATES

Closest to us on the evolutionary scale, many nonhuman primates demonstrate cognitive flexibility and beliefs.

• **Vervet Monkeys** The multiple, complex predator escape strategies adapted by these sub-Saharan monkeys reveal tremendously flexible calculations based on a wealth of knowledge and belief. How they escape and when they choose to bolt for cover (escape distance) is based on numerous factors, including their own speed and agility, their ability to self-camouflage in the terrain, the type, size and speed of the predator from eagles and snakes to leopards, past experience with the predator and the distance to refuge.

• **Orangutans** In one experiment, orangutans were presented with a clear tube with peanuts floating in water beyond their reach. No stones were available, but the animals were aware of drinking water in the next room. They went into the next room, filled their mouths with water, came back and spit a mouthful into the tube, raising the water level so that the peanuts floated up. They repeated this until they got all the peanuts. Then the experiment was repeated with an opaque tube so they couldn't see the peanuts. Nonetheless, based on the earlier experiment, they believed there were peanuts in the tube again, so they repeated their water-spitting strategy. Since they couldn't see the peanuts, they just kept going back for more water and spitting it into the tube until they got them. "This real-world problem-solving showed huge flexibility and knowledge of cause and effect. They appeared to know what would happen when they used the tools available to them (in this case water and their own mouths)," says Starzak. "I would say they demonstrated a definite causal belief system."

• **Chimpanzees** Chimps show great flexibility in problem-solving, adapting all manner of tools from their environment. For example, while young chimps don't know what on earth to do with a stick, older chimps have been observed finding twigs and long blades of grass, stripping the leaves off them, then finding a termite hole in the ground and inserting their newly fashioned fishing rod. When they pull out the tool, dozens of termites are

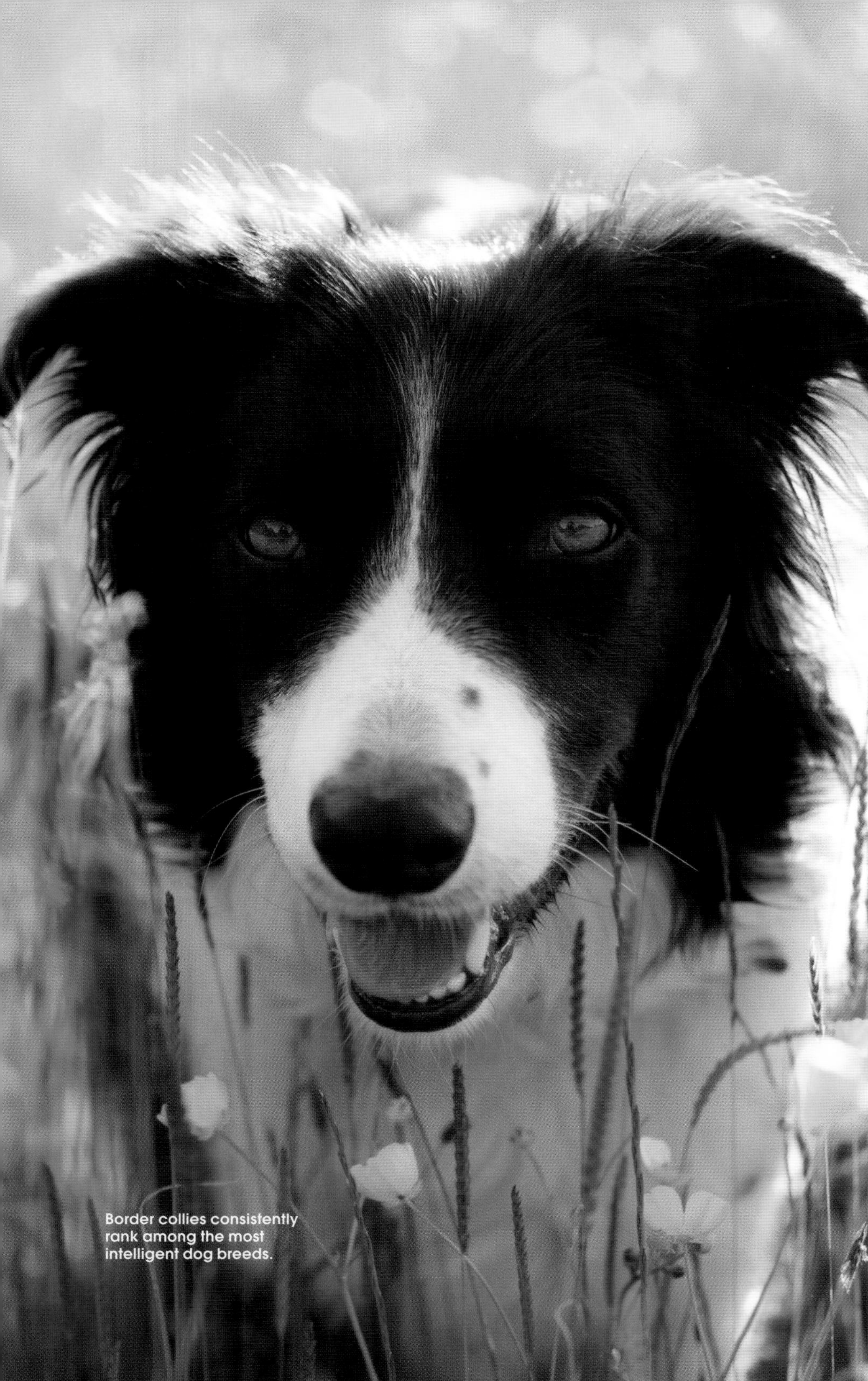

Border collies consistently rank among the most intelligent dog breeds.

crawling on it, a tremendous source of protein, and the chimps feast. Scientists literally call this behavior "termite fishing." It works with ants, too.

ASSUMING DOGS

In renowned experiments with the border collie Rico and other canines, dogs have shown they can associate many acoustic symbols (i.e., word labels/ names) with individual objects and fetch them on command. Rico knew the acoustic symbols for over 200 objects, and upon hearing a command such as "bring me Timi," he would fetch the right object from among the 200 on average 37 times out of 40. To do so, he learned to distinguish different shapes and sizes, and based on those qualities could group items into categories that helped him pinpoint the right objects.

Rico could learn a new object name in a single usage and respond to it correctly. Before he'd ever heard the word, the new object would be placed with seven familiar, named objects. Then using this new object's name, the researchers would ask him to retrieve it, and the dog would inevitably get it, presumably using a process of elimination.

"Dogs are very clever in some things and not so smart in others," explains Starzak. "They are very good at categorizing things." And that ability could not happen without their inherent belief in the permanent qualities of these objects. —*Mark Teich*

Mark Teich is an independent journalist living in New York City.

Blue-and-gold macaws form strong bonds with a mate and, when domesticated, with their human family.

School for Thought

Ecologist Carl Safina shares how different species raise their young, create beauty and achieve peace.

A sperm whale learns to embrace her fellow travelers. A macaw casts a covetous eye on his beautiful neighbor. A chimpanzee learns to pay to play. For ecologist and conservationist Carl Safina, humans are not the only ones with unique social cultures. In his latest book, *Becoming Wild: How Animal Cultures Raise Families, Create Beauty, and Achieve Peace*, he explores the cultural lives of sperm whales in the Caribbean, scarlet macaws in the Amazon forest of Peru and chimpanzees in Uganda's Budongo Forest.

Originally from Brooklyn, New York, Safina grew up in an apartment filled with singing canaries who were his father's hobby and passion. He spent weekends visiting the Bronx Zoo, the New York Aquarium, the American Museum of Natural History and aboard his uncle's boat. All this sparked a fascination with animals, and by age 7 he was raising homing pigeons. A love for camping in his teen years eventually lead to adventures in Kenya, Nepal, Greenland—and eventually to travels and investigations in all the continents and oceans of the Earth.

Safina earned a PhD from Rutgers University based on his study of seabirds. He then spent a decade overhauling fishing policies and restoring ocean wildlife. In the 1990s he led campaigns to ban high-seas drift nets that entrapped endangered fish and helped improve international management of fisheries targeting tunas and sharks. Along the way, he became a leading global voice for conservation.

Safina emphasizes that now, more than ever, protecting animals and their habitats matters. He wants us to appreciate the complexity and beauty of animal cultures in what he calls the "real world." No matter what human struggles we may be facing ourselves, in their other world, species with unique relationships, cultures and ways of teaching their young also are often struggling to survive.

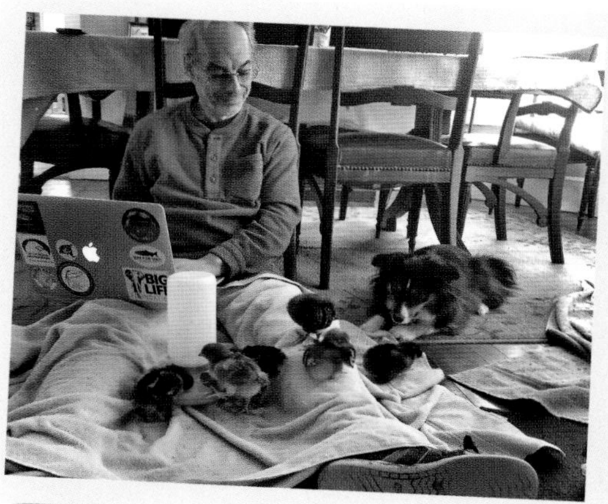

Conservationist Carl Safina works at home surrounded by some of his favorite friends.

What inspired *Becoming Wild*?
Writing my previous book, *Beyond Words*, led me to the question of culture and its evolutionary implications. I wanted to learn more about animal social groups and how they teach their young. I have a tremendous passion for observing animals in their habitats. Even in my own home, I love watching our animals and what they do. We have three dogs, nine laying hens, a rescued parrot, and in our yard lives an orphan hand-raised owl who found a mate and is about to hatch her first chicks. They are all beautiful, interesting and amusing. They teach the oneness of all living things, daily.

You suggest older animals train and teach their group's "culture" to younger members of the group.
Culture is what you learn socially about how and where you live, and can be passed along to subsequent generations. Take human language: We all learn to speak a language, but which one is purely cultural? Animal cultures vary widely depending on location and group. Their cultures are taught and learned.

You look primarily at sperm whales, chimpanzees and macaws. Why these animals in particular?
Sperm whales have a unique culture, as do chimpanzees. With wild macaws in Peru, I was very struck with the fact that they live in pairs. I wanted to uncover more about these relationships and the animals' interactions.

What surprised you the most about the sperm whales?
Sperm whales live in female-led groups that are organized like elephants. The males leave when they become adolescents. The females remain with their mother, aunts, sisters and babies. The reason for female-centered grouping is that their food source is often 2,000 feet below. Babies would not survive in those depths, so they wait at the surface with their babysitters, while their mothers dive below. The group is always in vocal contact. They live in families that are part of a clan.

Sperm whales have the ability to announce who they are as an individual, what family they belong to and what clan they belong to. They do so with a series of clicks that are like simple codes that are specific to the clan. The families within a clan get together, travel together and socialize together, but different clans avoid one another.

Sperm-whale clans may be made up of thousands of whales over thousands of square miles of ocean. They can tell by perceiving their vocal codes whether a whale is from their own clan or not.

Sperm whales live in female-led families that stay together and protect each other and their young for their entire lives.

Through vocal cues, sperm whales identify immediate family and more distant members of their clan.

Female chimpanzees are thought to be more trusting than males.

Grooming is a bonding experience for groups of chimpanzees and helps calm those who are nervous or tense.

There was a moving quote in *Becoming Wild* about how the discovery of whale song changed our understanding of animals.
Humpback whales sing elaborate songs; other whales have much simpler songs. Before the public heard the songs, they thought of whales as lumbering animals valued only for materials produced from their bodies. The discovery that these haunting sounds are actual songs—with patterns that the whales sing over and over again—changed how humans view whales. This change happened around the first Earth Day in 1970. When the people first heard recordings of whale songs, they burst into tears.

And what did you learn from observing macaws?
Macaws learn survival techniques from their parents. They also have pairing relationships. A pairing lasts throughout the nonbreeding season.

The discovery that whales produce beautiful and intricate songs propelled the Save the Whales movement.

They nest in the hollow of trees, which are in short supply. The pairs assess the ability of competing macaw pairs to obtain these nests. This assessing takes a lot of cognition. There is also cognition involved in their mate relationships. For example, there was a hawk headed straight for a macaws nest. The mate flew in, knocked the hawk sideways, and stopped it from causing harm.

You also write, "chimpanzees are… not half-baked humans," making the point that we are not superior to nonhuman animals.
Humans think that they are a perfect form of life. We look at other beings that are similar to us—apes and monkeys—and think "they are creatures in arrested development." That's wrong. Humans often act unwisely, stupidly, destructively. We pass our biases on to future generations. We often lack compassion and act with cruelty, enslave others, don't care for those in need. Humans have a long way to go.

Speaking of imperfections—you discuss the aggression of male chimpanzees, which you call a culturally learned trait. Are all male chimps violent?
In most areas and groupings, [chimpanzee] dominance is won through an aggressive physical contest. However, only a tiny percentage of the males contend for the top spot, and so many males don't engage in violence.

What about the literal act of murder among chimpanzees?
In many chimpanzee groups, physical violence and killing are part of life, similar to human societies. An animal will suddenly change his character and kill others inside his social group. Humans do the same thing. Very few other animal species behave this way.

Pay it forward: A baby humpback whale learns feeding techniques from peers and then passes that information on to others.

Are other primate species more reliably peaceful?
Bonobos are closely related to chimpanzees. They live in a totally different area. They never overlap. The bonobo females are the dominant figures, and they are peaceful. This peacefulness is common in many animal species where females are dominant.

Clearly, it's urgent that we take better care of the animals and the habitats in which they live.
Yes. Most living things are at their lowest population levels ever. We have taken over animal habitats, turned them into farmland and cities. Humans have polluted the Earth's water and air. We are acidifying the oceans. We are in the midst of an extinction crisis.

As the human population increases, so will demand for meat.

Cows can live naturally until about age 20, but many are slaughtered by age 5, after their third lactation cycle.

A Better Way *to Farm?*

Factory farming causes untold suffering. But in the future, meat grown as crops or cultivated in "breweries" will help us reduce animal agony—along with the threat of pandemics and climate change.

A glaring problem exists in factory farming, a leading cause of suffering and human public health and environmental risks. We know that factory farming is bad for animals. Every year, we breed, raise and kill more than 100 billion nonhuman creatures for food worldwide. Chickens, cows and pigs that are raised for meat in factory farms all suffer profoundly. We breed them to grow as big as possible as quickly as possible. We separate them from their families. We mutilate their bodies, typically without anesthesia. We force them to live in toxic, crowded buildings and transport them in hot, crowded trucks. And we slaughter them on disassembly lines that prioritize efficiency over reliability. While the details vary from case to case, industrial fish farming, industrial milk production, industrial egg production and other industrial animal agricultural systems all involve similar harms.

What few people know, but are now starting to realize, is that factory farming is bad for humans, too. There are many reasons why, ranging from its impact on workers to its impact on local communities, but two global issues are especially prevalent now. First, factory farming substantially increases the risk of pandemics, and—as the COVID-19 pandemic has shown—that's a risk none of us want to contribute to. For instance, industrial animal agriculture requires much more antimicrobial use than other industries, including human medicine. We use these drugs not only to stimulate growth, but also to control the spread of disease among farmed animals who are living in deeply toxic, crowded environments without access to veterinary care. As a result, factory farms are ideal breeding grounds for antimicrobial-resistant pathogens.

Hog, chicken and cattle waste has polluted 35,000 miles of rivers in 22 states and groundwater in 17 states.

The 2020 Australian bushfires have killed about one-third of all the koalas living in New South Wales, their main habitat.

Similarly, factory farming substantially increases the risk of climate change. It consumes much more land and water and produces much more waste and pollution than plant-based alternatives. This includes not only toxic waste but also food waste, since we need to grow many more plants to feed farmed animals than we would need to grow to nourish humans directly. Partly as a result of all of this consumption and pollution (and especially all of the deforestation that takes place to support it), industrial animal agriculture is a leading producer of greenhouse gases. In particular, this food system is responsible for an estimated 9% of carbon dioxide emissions, 37% of methane emissions and 65% of nitrous oxide emissions, adding up to an estimated 14.5% of greenhouse gas emissions over the next century.

When we learn about these harms, it can be easy to think: We should replace industrial animal agriculture with a nonindustrial alternative. That way we can preserve its good parts while reducing its bad parts. However, this solution has important limits. First, nonindustrial animal agriculture still harms and kills living things unnecessarily. Secondly, nonindustrial animal agriculture is even less sustainable than the industrial version, since it requires much more land per animal. Thus, even if nonindustrial animal agriculture were ethical, it is not scalable. While it might produce animal products for a relatively small number of wealthy people, it would not generate enough for a rising global population.

What this means is that, if we want to secure a better future for humans and other animals, then we must end industrial animal agriculture, substantially reducing meat production. At the same time, we must accept that nonindustrial animal agriculture can play, at most, a minor role in our future food system. In short, there is no path forward that will allow animal products to remain a central part of our lives. The only realistic options are a fully or mostly plant-based future in which nearly everyone is either fully or mostly vegan.

SCIENCE TO THE RESCUE

The good news is that with modern science, we'll still be eating what feels and tastes like meat! Many of us already have access to simple, healthful plant-based foods such as beans and lentils. But we might also crave food that plays the same role in our diets as meat, dairy and eggs. No worries. Researchers are now developing exactly that. Plant-based meat substitutes such as those produced by Beyond Meat and Impossible Foods are already widely available and popular among meat eaters and vegetarians alike. Additionally, cultivated meats—literal meat that we can produce without killing animals, by taking a cell culture from an animal and then using it to grow meat in a brewerylike factory—will soon be available, too. The more these technologies mature, the more we can all enjoy all the foods that we love while causing less harm to other humans, animals and the environment.

As we work toward this future, we can make social changes as well. Part of what makes an all plant-based diet hard for some people is that we were raised eating meat, and many of the cultural and religious traditions that we most deeply care about involve eating meat. This can make it seem like a plant-based future requires not only giving up a harmful food system, but also giving up a central part of who we are. But humans are adaptable, and history is a continual process of updating old traditions in light of new values. The more we reduce meat consumption, the more we can naturally reimagine shared traditions in more healthful, humane, sustainable ways. This will make it easier to preserve the parts of our identities that matter most to us while eliminating ones that cause harm.

We can also change laws in order to enable the safer, plant-based future we need. Part of what makes veganism hard for some people is that

All birds, including chickens, are excluded from federal animal protection laws in the United States.

The salmon in this catch could be carrying toxins that will make their way to the human dinner table.

meat seems so much more accessible and affordable than plant-based alternatives. This is partly because of taxes and other financial incentives. In the U.S., we subsidize the meat industry not only directly but indirectly by making the public pay for the health and environmental impacts of meat production. If we supported animal-based products less and plant foods more (or even if we simply reduced our support for the meat industry, and allowed animal and plant foods to compete on a level playing field), we would discover that plant foods are much more accessible and affordable overall.

With the right changes, a plant-based future is not only possible but desirable: We'll alleviate suffering for thousands of animals and also achieve greater health for ourselves. The catch is that we need to do more than simply wait for these changes to arrive. We must play an active role. We should try to reduce our meat, dairy and egg consumption as much as we can and also start talking about these issues in our everyday lives, as well as participate in local food movements and elections so that we can help bring about the social, political and economic changes that are necessary. None of these approaches is enough to bring about a more sustainable food system by itself. But if we combine such strategies in ways that make sense for us personally, we can make real progress. *—Jeff Sebo*

Jeff Sebo, PhD, is associate professor of environmental studies at New York University and co-author of Food, Animals, and the Environment *(2018).*

This whale shark filter-feeds in a polluted ocean, ingesting plastic that will make it ill.

Dirty Waters

Pollution can make some fish confused and passive, putting them at greater risk of predation and death.

When we humans ponder polluted waterways, our first thought tends to be the toxins we ourselves might consume when eating the local fish. But perhaps not surprisingly, polluted waters alter the fish's quality of life, too, as we now know them to be sensitive, intelligent and cognitively aware. Among the damages they incur: Polluted waters alter levels of the neurotransmitter serotonin (the same chemical altered by the antidepressant Prozac in humans), affecting such traits as aggression and boldness. Contaminated waters can also make fish more or less likely to explore the habitat, putting them at greater risk from predators. And organic pollutants such as pesticides cloud the cognitive ability of fish, making it more difficult for them to perceive alarm cues and learn.

Among the specific findings recently reported in the journal *Frontiers in Ecology and Evolution* by French researchers:

- When psychiatric drugs were dumped into a waterway, perch became more aggressive.
- When crucian carp were exposed to polystyrene nanoparticles, they seemed to have an altered brain structure that increased their feeding needs, thus exposing them to still more polystyrene and making them more voracious predators of other local species.
- Aluminum contamination impaired learning in Atlantic salmon that were challenged with passing through a maze. With lowered processing ability and impaired memory, they were poised to have trouble navigating new environments.
- Zebrafish that were contaminated with mercury became anxious and had trouble with locomotion.

These examples are disturbing, but not isolated. Researchers say that exposure to multiple pollutants and stressors, from predators to climate change, might now be the rule for fish in waters around the world. These chemical assaults can upend a fish species' emotional balance and cognition, damaging its fitness and ability to survive long term.

Pigs have unique personalities, a range of emotions and love to play.

Pigs are perhaps the most intelligent animal in the barnyard. They deserve the right to a living environment that meets their basic physical and psychological needs.

Fighting for Freedom

Do animals deserve their individual identities and rights?

"How should we relate to beings who look into mirrors and see themselves as individuals, who mourn companions and may die of grief, who have a consciousness of self? Don't they deserve to be treated with the same sort of consideration we accord to other highly sensitive beings: ourselves?" asks Jane Goodall.

In the fall of 2014, Steven Wise, founder and president of the Nonhuman Rights Project (NhRP), managed to peek inside an aluminum-sided shed near Gloversville, New York. A small television set cast enough of a glow for Wise to spot an adult chimpanzee crouched in the gloom. The great ape was locked behind a steel-mesh barrier, with only the TV and a few plastic toys for company. The shed belonged to a deceased man who had once been the owner of several circus chimps, according to a story in *The New York Times Magazine*. That man had raised this chimpanzee from infancy, and the ape's name was Tommy.

Tommy was about to make legal history—or so Wise hoped. After seeing Tommy in what Wise termed a "dungeon," the NhRP began appealing to courts to rule that this chimpanzee had the right to bodily liberty and should be released into a sanctuary to live among others of his own kind. They filed similar petitions on behalf of Kiko, another male chimpanzee, who is owned by a couple and apparently being held in isolation in a cage in a cement storefront in Niagara Falls, New York.

Given what scientists now know about our closest animal relatives—in brief, they are conscious, intelligent, emotional and highly social beings—it's doubtful that either Tommy or Kiko would choose the lives they are now living. But do they—or other great apes or dolphins, whales, elephants or African grey parrots—have the right to determine what happens to their lives? Wise and the NhRP argue that they do, that they should be considered persons with certain rights. But current United States law disagrees; it classifies an entity as either a person or a thing. Only persons have the capacity for rights, including being protected from unlawful confinement through the right of habeas corpus. Things do not have this capacity.

The Animal Legal Defense Fund's cases include a horse who sued his former abuser for the costs of his pain and suffering.

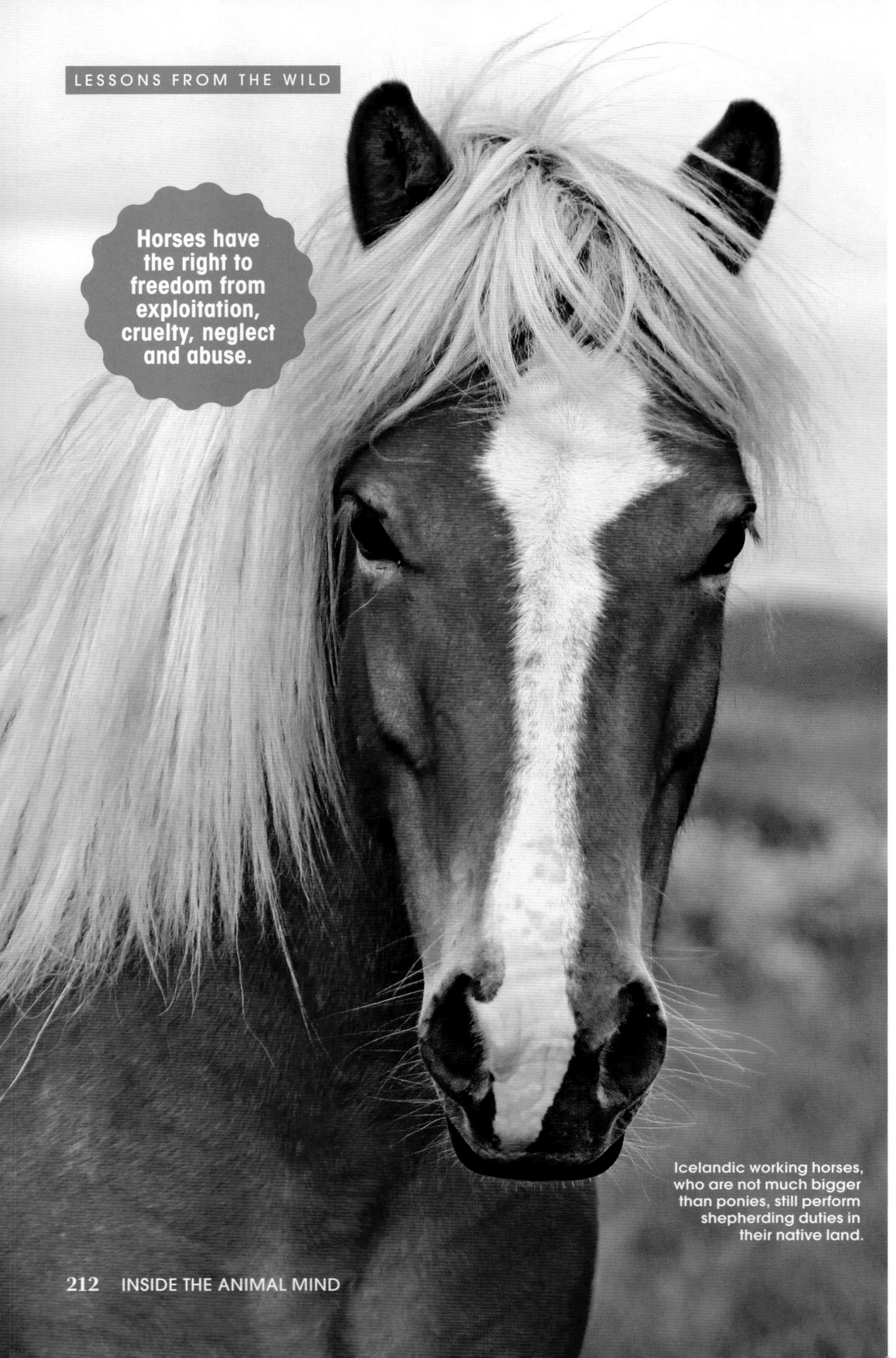
Horses have the right to freedom from exploitation, cruelty, neglect and abuse.

Icelandic working horses, who are not much bigger than ponies, still perform shepherding duties in their native land.

And sadly, despite all that science has discovered about the minds and emotions of animals in recent years, Tommy and Kiko (and all nonhuman animals, including the highly intelligent pig, the working horse and the large predators, such as lions, tigers and leopards) are considered things under federal law.

Still, there is the possibility that some animals, starting with the great apes, could be reclassified. After all, corporations as well as labor organizations, partnerships, associations and firms are all considered persons by the law because they have certain rights and duties and can sue or be sued.

The idea of animal personhood has gained ground elsewhere. In 2007, the parliament of the Balearic Islands, an autonomous community of Spain, passed the world's first law effectively granting legal personhood rights to all great apes. New Zealand has extended strong protections to the great-ape species, forbidding their use in medical research, testing or teaching—affording them what some scholars term "weak legal rights." Austria, the Netherlands and Sweden have also banned the use of great apes in animal testing.

In 2002, Germany amended its constitution, guaranteeing basic rights to all animals. India's Ministry of Environment, Forest and Climate Change banned the capture or importation of whales and dolphins for entertainment or exhibition purposes in 2013. The ministry also noted that many scientists regarded these animals as nonhuman persons, with commensurate rights—although this statement has no legal authority.

The chimpanzee Cecilia (above) was declared a "nonhuman legal person" by an Argentine judge in 2016.

Most promising, in November 2016, an Argentine judge declared Cecilia, a chimpanzee, a "nonhuman legal person" with "inherent rights"—exactly what Wise and NhRP have been fighting for on behalf of Tommy and Kiko. The landmark ruling freed Cecilia, who had been caged alone for years in the Mendoza Zoo, and cleared the way for her departure to the Great Ape Project's sanctuary in Brazil—where she now walks on grass and has other chimpanzees for companions.

The ruling, sought by the Association of Professional Lawyers for Animal Rights, marks the first time in the world's judicial history that an animal has benefited from the right of habeas corpus. Judge María Alejandra Mauricio told the newspaper *Los Andes*, "We're not talking about civil rights enshrined in the Civil Code. We're talking about the species' own rights: development and life in their natural habitat."

But the Nonhuman Rights Project in the U.S. has so far failed to persuade courts to acknowledge the personhood of Tommy and Kiko—although they have received support from many scientists. In 2018, 17 philosophers from leading universities submitted an amicus curiae brief to the New York State Court of Appeals on behalf of the chimpanzees. Jeff Sebo, director of animal studies at New York University, was part of the group. "The idea of nonhuman personhood," he wrote, "might seem confusing at first, since we tend to use the terms 'human' and 'person' interchangeably. But they are not equivalent."

Human is a "biological concept," referring to a *Homo sapiens*. *Person* refers to an individual who can hold moral and legal rights. If Tommy and Kiko are persons, what rights can they have? If they have the right to liberty, Sebo says, do they also have the right to property or free expression? And if these chimpanzees have rights, what about cats, dogs, pigs, cows or even ants? Where do we draw the line?

Society and the courts are struggling to answer these unsettling questions.

Do animals deserve the right to freedom of movement or the ability to socialize with others of their kind?

Animal Bill of Rights

The Animal Legal Defense Fund has developed an Animal Bill of Rights, which exists as a petition to the United States Congress. These basic, inalienable rights, according to the fund, should apply to all sentient beings.

I. The right of animals to be free from exploitation, cruelty, neglect and abuse.

II. The right of laboratory animals not to be used in cruel or unnecessary experiments.

III. The right of companion animals to a healthy diet, protective shelter and adequate medical care.

IV. The right of wildlife to a natural habitat that is ecologically sufficient to a normal existence and a self-sustaining population.

V. The right of farmed animals to an environment that satisfies their basic physical and psychological needs.

VI. The right of animals to have their interests represented in court and safeguarded by the law of the land.

In a description of their brief, two of the philosophers, Andrew Fenton and Syd M. Johnson, point out why they feel the issue are important: "Why does it matter if the courts agree that Tommy and Kiko are persons? Right now, they are being held in solitary enclosures and are legally unprotected from being confined in this way, even though doing so harms them. They are unprotected because under the law, there are only persons or things, and Tommy and Kiko are not recognized as persons. As far as we know, their current owners are not breaking any animal welfare laws, so their solitary captivity seems perfectly legal.

"This is a fundamental flaw with animal-welfare laws—while they can protect animals from some forms of outrageous abuse (starvation, neglect), they are otherwise silent about whether other important interests of animals are served, such as their freedom of movement or ability to socialize with others of their kind. Only persons have rights, including rights to bodily liberty. So, unless and until Tommy and Kiko are recognized as persons, they remain legal things, lacking even the most basic rights, including the right to live as chimpanzees, with other chimpanzees."

In 2018, the New York State Court of Appeals denied the NhRP's motion to appeal a lower court's ruling against the habeas corpus petition. There was no legal precedent for chimpanzees to be considered persons, the court ruled; nor can chimpanzees be held legally accountable for their actions.

One of the judges, Eugene M. Fahey, issued a concurring opinion, asking "whether a chimpanzee is an individual, with inherent value who has the right to be treated with respect." He added a personal reflection, acknowledging his struggle over the issue of personhood for chimpanzees and other animals. "The issue whether a nonhuman animal has a fundamental right to liberty protected by the writ of habeas corpus is profound and far-reaching," Fahey wrote. "It speaks to our relationship with all the life around us. Ultimately, we will not be able to ignore it. While it may be arguable that a chimpanzee is not a 'person,' there is no doubt that it is not merely a thing."

Wise is not giving up. "For 2,000 years, all nonhuman animals have been legal things who lack the capacity for any legal rights. This is not going to change without a struggle." —*Virginia Morell*

Virginia Morell is the author of Becoming a Marine Biologist *and* Animal Wise.

Did You Know?
Gorillas are so intelligent, they can sign hundreds or thousands of words, understand human languages and laugh at their own jokes and the jokes of others. They can even lie in order to get out of trouble.

Leopards deserve a natural habitat that supports a self-sustaining population.

A leopard rests on a tree at the Masai Mara National Reserve for game in Kenya.

Fun Facts About the Animal Mind

Elephants can count. Dogs read our expressions. Cows have personality, and fish feel pain. New science across the species keeps rolling in.

1

In addition to male breeding calls, **humpback whales** of both sexes communicate through a plethora of sounds, from whoops to trumpets to growls. Humpback whales in the North Pacific can pass on distinctive calls through three generations.

2

Dogs can read feelings on people's faces. When shown a photo of an angry, fearful or happy face, dogs tend to turn their heads to the left, while surprised faces elicit a rightward turn. Experts speculate that canines process different emotions using different hemispheres of the brain. Expressions of anger, fear and happiness also trigger signs of stress in dogs.

3

Male smooth guardian frogs in Borneo will sit virtually motionless for days on end while caring for a clutch of eggs. Females of the species, meanwhile, compete among themselves for attention from the opposite sex, initiating mating calls to males.

4

Asian elephants have the most advanced counting skills of any animal, save humans. In one study, an elephant was trained to use a touch screen by tapping it with her trunk. When presented with two images of fruit clusters, the numerically savvy animal was able to choose the image with the larger number of items two-thirds of the time. Another Asian elephant, called Happy, became the first of her species in the world to pass the mirror test of self-recognition.

5

Termites are the world's greatest insect engineers. In a forest in northeastern Brazil, scientists have recently discovered a vast array of some 200 million dirt mounds covering an area the size of Great Britain. It's said that an ancient civilization of termites piled up the conical heaps of soil while carving out sprawling underground tunnels approximately 4,000 years ago.

6

Chimpanzees may be the only nonhuman creatures to understand death is irreversible. For example, chimps in the Ivory Coast have been observed licking the wounds of injured—but not dead—members of their species. And when a young chimp was killed after falling from a tree in Tanzania, 16 chimps from the victim's group commenced "raucous calling...slapping and stamping the ground, tearing and dragging vegetation, and throwing large stones," said an anthropologist who studied the incident.

7

Ravens gesture with their beaks and wings, using the appendages to communicate much as humans use their hands. Researchers have observed wild ravens wielding their beaks to point at items such as moss and twigs, often prompting members of the opposite sex to look at the objects and engage with the gesticulating bird. These types of referential gestures are almost unheard of among nonhuman animals. Even primates don't often point.

8

Fish feel pain, according to surprising—and growing—scientific consensus. Fish have sensory neurons called nociceptors that detect painful stimuli. When the aquatic animals are injured, their brains light up in ways that suggest they can feel harm rather than just react to it. Fish also exhibit unusual, distracted behaviors when injected with painful acetic acid, suggesting a conscious experience of suffering. The once-common view that fish brains aren't complex enough to experience pain has given way to a recognition that, like mammals and birds, fish, too, can hurt.

9

Young males in many **songbird** species keep their feminine plumage colors long after they mature. Scientists believe they do this to pose as female and therefore avoid being targeted for aggression by older male birds. Usually the deception is temporary, but in some species, males make a permanent switch: In one bird of prey, some 40% of males pose as females to gain access to other males' territories.

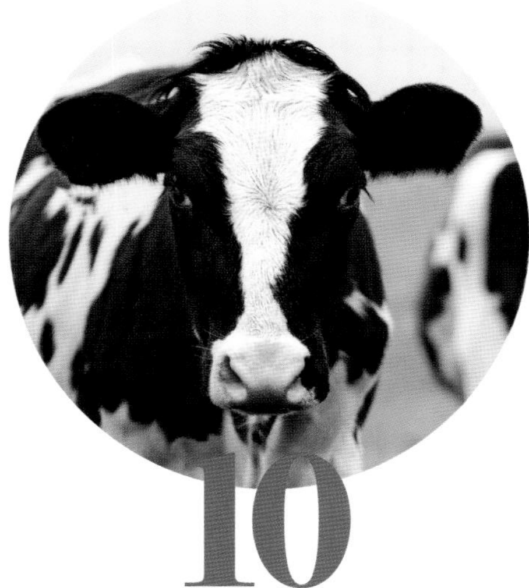

10

Cows have personality traits. Dairy calves, for example, show pessimism, fearfulness and sociability from a very young age. According to studies, these traits also predict the animals' ability to cope with stress. Gloomier calves have more intense emotional responses to challenging situations, such as being transported between barns. Their eye temperature rises and they vocalize more often than their brethren, signaling that they are more stressed out.

SPECIAL THANKS TO CONTRIBUTING WRITERS

Sherry Baker • Erica Cirino • Payal Dhar
Temma Ehrenfeld • Samantha Facciala • Katharine Gammon
Amanda Giracca • Sharon Guynup • Keith Harary
Michele C. Hollow • Heidi Hutner • Gregory Isaacson
Lori Marino • Virginia Morell • April Reese
Anita Salzberg • Jeff Sebo • Heather Swan
Jason Teich • Mark Teich • Kaitlin Stack Whitney
Emily Willingham

CREDITS

COVER Faba-Photography/Getty Images; RichLegg/Getty Images; Hung Chung Chih/Shutterstock; Nimit Virdi / 500px/Getty Images; Tier Und Naturfotografie J und C Sohns/Getty Images; Steffen Benz/EyeEm/Getty Images; Bobiko/Getty Images;Luca Lopez/EyeEm/Getty Images **FRONT FLAP** 1001slide/Getty Images **2-3** moggara12/Shutterstock **4-5** chendongshan/Getty Images; sethakan/Getty Images; Tier Und Naturfotografie Jund C Sohns/Getty Images; pchoui/Getty Images **6-7** Kondratuk Aleksei/Shutterstock **8-9** Hero Images/Getty Images **10-11** Emely/Getty Images **12-13** Ryan Cutler/EyeEm **14-15** Marcia Fernandes/Getty Images; Kevin Vandenberghe/Getty Images **16-17** Purple Collar Pet Photography/Getty Images; Jagodka/Shutterstock **18-19** Westend61/Getty Images **20-21** Marcel ter Bekke/Getty Images; Margo Brosko/Shutterstock; qbanczyk/Getty Images **22-23** Westhoff/Getty Images; Seregraff/Shutterstock **24-25** Nina Goldenberg/EyeEm/Getty Images **26-27** Luxx Images/Getty Images; Caiaimage/Getty Images; Agency Animal Picture/Getty Images; SensorSpot/Getty Images; Dora Zett/Shutterstock_; Anthony Lee/Getty Images **28-29** chendongshan/Getty Images; Hans Surfer/Getty Images; Life On White/Getty Images **30-31** PeopleImages/Getty Images; EyeEm/Getty Images **32-33** Javier Brosch/Shutterstock **34-35** caroldejorge.fotografia/Shutterstock **36-37** Jeffrey Coolidge/Getty Images **38-39** Purple Collar Pet Photography/Getty Images; Sol de Zuasnabar Brebbia/Getty Images **40-41** Sabine Arnold/EyeEm/Getty Images **42-43** Holly Michele/Shutterstock; Juan Carlos Vindas/Getty Images **44-45** mehmettorlak/Getty Images **46-47** Cavan Images/Getty Images; Tony Stock/Shutterstock **48-49** Georgette Douwma/Getty Images **50-51** Westend61/Getty Images **52-53** JIANG HONGYAN/Shutterstock; Westend61/Getty Images **54-55** Chendongshan/ Shutterstock **56-57** Ann Schwede/Getty Images; Monkey Business Images/Shutterstock **58-59** MintImages/Shutterstock **60-61** cynoclub/Getty Images; Olhastock/Shutterstock; Nynke van Holten/Getty Images **62-63** mantaphoto/Getty Images **64-65** Alan Tunnicliffe Photography/Getty Images **66-67** Oliver Rossi/Getty Images; xbrchx/Shutterstock; Ery Azmeer/Shutterstock; Shinedawn/Shutterstock; TreesTons/Shutterstock; DWI YULIANTO/Shutterstock; Dobermaraner/Shutterstock **68-69** Stephen Kenneston **70-71** Stephen Kenneston **72-73** narvikk/Getty Images **74-75** Martin Ruegner/Getty Images **76-77** SiiKA Photo/Shutterstock; Willyam Bradberry/Shutterstock **78-79** slowmotiongli/Shutterstock; Katy Laveck Foster/NOAA Fisheries **80-81** sethakan/Getty Images **82-83** Thanaphong Araveeporn/Shutterstock **84-85** THEPALMER/Getty Images **86-87** WLDavies/Getty Images; Robin Wilson Photography/Getty Images **88-89** spxChrome/Getty Images; 1001slide/Getty Images **90-91** ArtMarie/ Getty Images **92-93** Antagain/Getty Images; BrilliantEye/Getty Images **94-95** svetikd/Getty Images; David Lazar/Getty Images **96-97** TheDman/Getty Images **98-99** Sarah Cheriton-Jones/Getty Images; slowmotiongli/Getty Images **100-101** Rüdiger Katterwe/EyeEm/Getty Images **102-103** Daniel Hernanz Ramos/Getty Images; madcorona/Getty Images **104-105** David B Waters/500px/Getty Images **106-107** Mia_Garrett/Getty Images **108-109** Philippe LEJEANVRE/Getty Images **110-111** Mint Images/Getty Images; Antagain/Getty Images; Jacky Parker Photography/Getty Images **112-113** Frédéric Collin/Getty Images; Antagain/Getty Images **114-115** Anup Shah/Getty Images; CBS Photo Archive/Getty Images **116-117** Picture by Tambako the Jaguar/Getty Images; CBS Photo Archive/Getty Images **118-119** Anup Shah/Getty Images; Apic/Getty Images **120-121** rogertrentham/ Getty Images **122-123** dageldog/Getty Images; your personal camera obscura/Getty Images **124-125** Geraint Rowland Photography/Getty Images; SolStock/Getty Images **126-127** Marc Guitard/Getty Images **128-129** georgeclerk/Getty Images **130-131** SDI Productions/Getty Images; Johner Images/ Getty Images **132-133** kupicoo/Getty Images; georgeclerk/Getty Images **134-135** David Rasmus/Shutterstock **136-137** Dev_Maryna/Shutterstock; LorenaKaz **138-139** Manoj Shah/Getty Images; Thomas Foldes/Shutterstock; Vladimir Wrangel/Shutterstock; Michaela Walch/imageBROKER/ Shutterstock **140-141** Andrea Izzotti/Shutterstock **142-143** R. Andrew Odum/Getty Images; Peter Weber/Getty Images **144-145** Nathapol Kongseang/ Shutterstock; Vladimir Wrangel/Shutterstock **146-147** Tetra Images/Getty Images; photomaster/Shutterstock **148-149** Agency Animal Picture/Getty Images; Barbara Rich/Getty Images **150-151** eyepix/Getty Images; Perkus/Getty Images **152-153** Compassionate Eye Foundation/David Leahy/Getty Images **154-155** TerryJ/Getty Images **156-157** David Osberg/Getty Images; Kennan Harvey/Getty Images **158-159** BRENDAN SMIALOWSKI/Getty Images ; Stocktrek Images/Getty Images; Elles Rijsdijk/EyeEm/Getty Images **160-161** Tier Und Naturfotografie J und C Sohns/Getty Images **162-163** mikdam/Getty Images; Rick Friedman/Getty Images **164-165** tap10/Getty Images; Debi Dalio/Getty Images **166-167** TommL/Getty Images; KIM JAE-HWAN /Getty Images **168-169** Guntaphons Sombutniranart /EyeEm/Getty Images **170-171** WikiCommons **172-173** Courtesy of the Houston Zoo; Universal History Archive/Getty Images (2) **174-175** Sajeev Krishnan / 500px/Getty Images **176-177** Sean Gallup/Getty Images **178-179** Sean Gallup/Getty Images; lux3000/Shutterstock **180-181** Sean Gallup/Getty Images **182-183** Aliaksandr Sapeha/Getty Images; Sergey Sosnitsky/ Shutterstock; Kate Siomkina/Shutterstock **184-185** Bernard Radvaner/Getty Images **186-187** Reynold Mainse/Design Pics/Getty Images; Image Source/Getty Images **188-189** pr2is/Shutterstock; Oksana Maksymova/Shutterstock **190-191** Busakorn Pongparnit/Getty Images; Carl Safina **192-193** wildestanimal/Getty Images **194-195** Ari Wid/Shutterstock; Craig Lambert Photography/Shutterstock **196-197** Anup Shah/Getty Images **198-199** KenCanning/Getty Images **200-201** Bart Groundhog/Shutterstock; Marcin Perkowski/Shutterstock **202-203** Photos by R A Kearton/Getty Images **204-205** Vicki Smith/Getty Images **206-207** Felix Cesare/Getty Images; Nay Wachr Phngs Kaew Thùng/EyeEm/Getty Images **208-209** bksrus/ Getty Images; Rich Carey/Shutterstock **210-211** Bruno Raffa/EyeEm/Getty Images **212-213** Gigja Einarsdottir/Getty Images; AFP Contributor/Getty Images **214-215** Chris Clor/Getty Images; pchoui/Getty Images **216-217** wildestanimal/Getty Images; Image Quest Marine/Alamy Stock Photo; 500px/ Getty Images; Kyrin Geisser/EyeEm/Getty Images; Cappi Thompson/Getty Images; Faba-Photograhpy/Getty Images **BACK FLAP** mehmettorlak/Getty Images; spxChrome/Gett Images **BACK COVER** JIANG HONGYAN/Shutterstock; Olhastock/Shutterstock; Aksenova Natalya/Shutterstock; sethakan/ Getty Images; Vicki Smith/Getty Images; kamnuan/Shutterstock

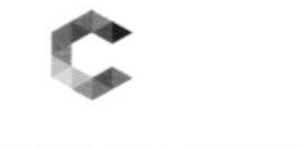

CENTENNIAL BOOKS

An Imprint of
Centennial Media, LLC
40 Worth St., 10th Floor
New York, NY 10013, U.S.A.

CENTENNIAL BOOKS is a trademark of Centennial Media, LLC

ISBN 978-1-951274-61-0

Distributed by
Simon & Schuster, Inc.
1230 Avenue of the Americas
New York, NY 10020, U.S.A.

For information about custom editions, special sales and premium and corporate purchases, please contact Centennial Media at contact@centennialmedia.com.

Manufactured in China

10 9 8 7 6 5 4 3 2 1

Publishers & Co-Founders Ben Harris, Sebastian Raatz
Editorial Director Annabel Vered
Creative Director Jessica Power
Executive Editor Janet Giovanelli
Features Editor Alyssa Shaffer
Deputy Editors Ron Kelly, Anne Marie O'Connor
Design Director Martin Elfers
Senior Art Director Pino Impastato
Art Directors Patrick Crowley, Natali Suasnavas, Joseph Ulatowski
Copy/Production Patty Carroll, Angela Taormina
Assistant Art Director Jaclyn Loney
Photo Editor Jenny Veiga
Production Manager Paul Rodina
Production Assistant Alyssa Swiderski
Editorial Assistant Tiana Schippa
Sales & Marketing Jeremy Nurnberg